GLOBAL SCIENCE & TECHNOLOGY INFORMATION

A New Spin on Access

Caroline S. Wagner · Allison Yezril

Supported by the National Science Foundation

Science and Technology Policy Institute

RAND

Library of Congress Cataloging-in-Publication Data

Wagner, Caroline S.
      Global science & technology information : a new spin on access/
Caroline S. Wagner and Allison Yezril.
            p.   cm.
      "Supported by the National Science Foundation. Prepared by
RAND's Science and Technology Policy Institute."
      "MR-1079-NSF."
      Includes bibliographical references.
      ISBN 0-8330-2755-7
      1.   Science—United States—Information services—Evaluation.
2. Technology—United States—Information services—Evaluation.
3. Government information agencies—United States—Evaluation.
I. Yezril, Allison.   II. Title.   III. Title: Global science and
technology information.
Q224.3.U6W34   1999
338.97306—dc21   99-38209
                                                                CIP

Published 1999 by RAND
1700 Main Street, P.O. Box 2138, Santa Monica, CA 90407-2138
1333 H St., N.W., Washington, D.C. 20005-4707
RAND URL: http://www.rand.org/
To order RAND documents or to obtain additional information, contact
Distribution Services: Telephone: (310) 451-7002; Fax: (310) 451-6915;
Internet: order@rand.org

This report provides findings from a project conducted by the Science and Technology Policy Institute (S&TPI) at RAND for the National Science Foundation (NSF). NSF requested a report on the current and future need for global technology assessment and science and technology (S&T) information among government and industry decisionmakers and researchers. In particular, NSF asked S&TPI to examine the role of the World Technology Evaluation Center (WTEC) in providing global technology assessment services in the future. Paul Herer, Engineering Directorate, NSF, commissioned the study.

Created by Congress in 1991, the Critical Technologies Institute was renamed in 1998 as the Science and Technology Policy Institute. The Institute is a federally funded research and development center sponsored by the National Science Foundation and managed by RAND. The Institute's mission is to help improve public policy by conducting objective, independent research and analysis on policy issues that involve science and technology. To this end, the Institute

- supports the Office of Science and Technology Policy and other Executive Branch agencies, offices, and councils

- helps science and technology decisionmakers understand the likely consequences of their decisions and choose among alternative policies

- helps improve understanding in both the public and private sectors of the ways in which science and technology can better serve national objectives.

S&T Policy Institute research focuses on problems of science and technology policy that involve multiple agencies. In carrying out its mission, the Institute consults broadly with representatives from private industry, institutions of higher education, and other nonprofit institutions.

Inquiries regarding the S&T Policy Institute or this document may be directed to:

Bruce Don
Director, Science and Technology Policy Institute

RAND
1333 H St., N.W.
Washington, D.C. 20005

Phone: (202) 296-5000
Web: http://www.rand.org/center/stpi
Email: stpi@rand.org

CONTENTS

FIGURE AND TABLES

Figure

Table

As our trade partners become increasingly sophisticated, we must look outward for knowledge, not just for new markets. Useful ideas will percolate in unusual places. Far from being a process to be resisted, the globalization of industrial R&D can benefit us if we learn to learn.

—*Gary C. Hufbauer (Callan et al., 1999, Foreword).*

Global technology assessment and programs providing international science and technology information (ISTI) have been valued government services and continue to be needed by both government and industry. This is the consensus of respondents to a RAND survey of more than 90 users of science and technology information. Overwhelmingly, survey respondents said they will continue to need global technology assessment and foreign S&T information in the future. Moreover, the majority surveyed said that ISTI collection and analysis remain important roles for government. More respondents thought that a mix of government and private-sector sources was more likely to meet their needs than either private or government services alone. These respondents said that, while private sources are often more in-depth and strategic in approach, government sources are less biased and provide a broad baseline of information.

AN EVOLVING GOVERNMENT ROLE IN PROVIDING ISTI

The U.S. federal government has been tracking and monitoring ISTI for more than three decades. During that time, the global marketplace and the economy have changed dramatically. S&T capabilities in other nations have matured, some to world-class levels, resulting in highly competitive technology-based products. As factors have changed and S&T capabilities have dispersed, the types of ISTI information U.S. entities need have also changed. Government ISTI services have adapted to these changes. We identified two historical transitions that have occurred in ISTI services in response to external changes: one a postwar shift from defense-oriented analysis to include civilian S&T, and the other, intensified reporting on Japan and Germany in response to 1970s trade competition.

The first of the transitions for ISTI services occurred after the Second World War, when it became apparent that understanding scientific and technical developments in key countries would aid general welfare and national security. The Office of Scientific Research and Development (OSRD) had proved highly effective during the war in identifying important S&T developments that aided the war effort. A postwar function incorporating some of OSRD's methods of technology assessment that would bridge military and civilian S&T was suggested. In recommendations that eventually led to the creation of the NSF, Vannevar Bush, the wartime director of OSRD, said that the science agency should "promote the dissemination of scientific and technical information and . . . further its international exchange." (Bush, 1960, p. 37, item m.) As a result, scientific and technological intelligence was initiated in U.S. government intelligence and scientific agencies.

The second transition in government ISTI services came in the 1970s, when the "Japanese challenge" raised awareness of the increasingly competitive nature of technology being developed abroad. Specifically, during the 1970s through the 1980s, there was significant alarm about Japanese technological ascendancy. As a result, dozens of ISTI programs were created in seven or eight government agencies (GAO, 1993). These included programs to provide

- literature reviews and monitoring, such as the Japanese Technical Literature Program
- "snapshot" assessments of foreign technological capabilities, such as the Science and Technology Reporting Information Dissemination and Enhancement (STRIDE) program
- analytic reports on foreign technical capabilities, such as WTEC.

Government agencies also collected ISTI for their own use; this information was sometimes shared across the government. Other private and nonprofit groups, such as SRI International and Price Waterhouse, also began conducting similar analyses.

DETERMINING GLOBAL ASSESSMENT NEEDS FOR THE FUTURE

Today, the shifting landscape for global S&T suggests that demand among business and government users for ISTI is in the process of a third transition. This transition is being effected by (1) the rapid globalization of the world economy, (2) the shifting perceptions of the competitive posture of U.S. technological strength, and (3) the changes in information collection and dissemination being fostered by the information revolution. As in earlier transitions, this has implications for the ways in which the U.S. government collects, analyzes, and disseminates ISTI.

In an effort to garner opinions and assess demand for ISTI, RAND asked a group of researchers and policymakers how and why they use ISTI and what kinds of information they will need in the future. Overwhelmingly, respondents said that ISTI collection and analysis are important roles for government. Moreover, nearly all of those interviewed said they would continue to need, and increasingly require, ISTI in the future. Respondents also agreed that global technology assessment should expand beyond Japan and Europe to encompass global developments. In fact, most respondents were not as interested in a *country* focus as they were in a *technology* focus.

While most respondents reported using a range of public and private sources of information, they found government-funded sources to be more objective and trustworthy than private sources. A significant number of those contacted reported that the lower cost of government services was a factor in their use of government ISTI services. The government's access to vast information resources around the world was also cited as a reason government should continue and expand ISTI services.

While existing government services were generally judged useful, they came in for three kinds of criticisms: (1) Government information is too broad or not timely enough; (2) technical information is provided without adequate political or economic context; and (3) the information is neither well-coordinated across the government nor easy to access.

OPTIONS FOR RESPONDING TO THE THIRD TRANSITION

Users today need information more quickly, from more sources and geographic regions, and within a richer political and economic context than in the past. New information tools can be used in at least four ways to meet these emerging needs. One, use Internet search capabilities to conduct "smart" searches of bibliographic information for the individual user. Two, apply software systems to query various databases and provide coordinated responses without having to build a new database. Three, use information-management tools to hold "virtual meetings" or surveys of experts on a periodic or continuous basis. Finally, parse information in a "tiered system" to meet the needs of the casual, as well as the highly technical, user.

ACCOUNTING FOR SUCCESS

As government seeks to improve access to ISTI services, new measures of effectiveness will need to be developed. In the era of accountability created by the Government Performance and Results Act, government program managers are required to set goals and account for the outputs and outcomes of government activities. As government agencies explore ways to meet the needs of the third transition user for global S&T information, goals for improvement must be set, and measures must be developed to assess progress toward meeting these goals. Potential improvements to ISTI activities fall into five categories:

1. efficiency of information gathering and production

2. quality of the verbal and written reports

3. timeliness of information delivery

4. accessibility of information

5. impact of information.

Both qualitative and quantitative measures, when placed in context, can provide useful measures of the extent to which program goals are being met.

Quantitative measures—placing numerical or "counted" values on the outcomes of program activities—while more controversial than qualitative measures, are also more compelling and politically acceptable. Quantitative measures that could be applied to ISTI services are input-output measures, citation counts, timeliness measures, market tests of quality ("willingness to pay"), and counts of increasing numbers of users. To be most effective, quantitative information must be married to qualitative information and placed in the context of the full program's goals and activities.

Qualitative measures, such as peer-review, expert judgment, and user surveys, remain effective measures of the usefulness of information-based services. Qualitative measures for assessing the creation of new knowledge have been heavily, indeed almost exclusively, used to assess knowledge-generating activities. They are highly developed and easy to use. Nevertheless, these measures have some drawbacks with relation to better program accountability. They are criticized as focusing more heavily on inputs than on outputs and as lacking objectivity and rigor. Qualitative measures, when complemented by quantitative measures, can provide a fuller, more robust picture of program effectiveness

CONCLUSION

Our survey suggests that there is a need for timely, unbiased, and global S&T information that is easily accessible to users and that provides a global picture of S&T developments. The public need for this information is growing, and a continued and possibly expanded role for government appears appropriate. Expanding the mission of the government's activities in assessment to include the following would greatly enhance and update services and meet the needs of government and private-sector users:

1. providing a global overview

2. including economic and trade information to place technologies in context

3. encouraging the networking of information and resources across the government and possibly with other information sources

4. increasing the number of quick responses and person-to-person interactions to share technical know-how.

Many of these improvements can be made using the very factors that are shifting the landscape for global technology assessment: the globalization of industry, the information revolution, and dispersed scientific excellence.

- The business network that results from globalization provides access to a broad range of globally dispersed experts.

- Electronic information management tools emerging from the information revolution provide the opportunity for ISTI services to access vastly more information and present it more efficiently.

- Dispersed scientific excellence offers the chance to leverage investments in research and development.

Using new electronic tools and methods would not preclude the need for expert judgment, on-site visits, and workshops such as the ones WTEC has conducted effectively. Rather, the effective features of existing services, combined with new information tools and opportunities, can be incorporated to build the next generation of ISTI services.

The authors wish to thank Melissa Bradley, a RAND survey research specialist, who helped design the interview tool used for this study and who also conducted a number of the interviews. Paul Steinberg, a RAND communications analyst, was of considerable assistance in advising on issues of presentation and organization for the report.

The management and staff of the International Technology Research Institute at Loyola College in Maryland provided background information and data that proved highly useful in conducting this study.

Department of Commerce officials also provided both data and insight that assisted this project. Particular thanks go to Phyllis Genther-Yoshida (Technology Administration, Department of Commerce), Patricia McNutt (National Technical Information Service), and Abram Shulsky (RAND).

Finally, the authors wish to thank those individuals in the private-sector firms and government offices with whom we conducted interviews. Their ideas and responses greatly enriched the information available in this report.

ABBREVIATIONS

ARO	Army Research Office
AFSC	Air Force Systems Command
ATIP	Asian Technology Information Program
CIA	Central Intelligence Agency
CISAR	Community-Initiated State of the Art Reviews
COSATI	Committee of Science and Technical Information
DARPA	Defense Advanced Research Projects Agency
DoC	Department of Commerce
DoE	Department of Energy
FBIS	Foreign Broadcast Information Service
GAO	U.S. General Accounting Office
GPRA	Government Performance and Results Act of 1993
ISTI	International science and technology information
ITRI	International Technology Research Institute
JETRO	Japan External Trade Organization
JTECH	Japanese Technology Evaluation Program
JTEC	Japanese Technology Evaluation Center
JTLP	Japanese Technical Literature Program
MCC	Microelectronics and Computer Technology Corporation
MEMS	Micro Electro-Mechanical Systems
MIT	Massachusetts Institute of Technology
MITI	Japanese Ministry of International Trade and Industry
NASA	National Aeronautics and Space Administration
NIH	National Institutes of Health

NIST   National Institute of Standards and Technology
NSF   National Science Foundation
NTIS   National Technical Information Service
ONR   Office of Naval Research
OSR   Air Force Office of Scientific Research
OSRD   Office of Scientific Research and Development
OSTP   Office of Science and Technology Policy
OTA   Office of Technology Assessment
R&D   Research and development
S&E   Scientists and engineers
S&T   Science and technology
S&TPI   Science and Technology Policy Institute
SRI   Stanford Research International
STRIDE   Science and Technology Reporting Information
                Dissemination and Enhancement
WTEC   World Technology Evaluation Center

Chapter One

INTRODUCTION

OBJECTIVE AND APPROACH

This report responds to a request the National Science Foundation (NSF) made to RAND to inquire into current and future needs for global technology assessment and science and technology (S&T) information among government and industry decisionmakers and researchers. Specifically, this project set out to answer the following questions:

- What public needs are being met with global technology assessment products, and what is the federal role in supporting this?

- Who are the current and potential customers?

- Can and should major improvements be made in current study methodology and dissemination activities that would better meet government mission or customer demand?

- How can these services measure performance over time and can these measures be built into activities in a way that provides continual feedback?

To answer these questions, the project conducted a review of the existing literature to develop an historical context for understanding the current structure of U.S. government services providing global technology assessment. In addition, the project team developed a list of questions and conducted a series of interviews to elicit information from policymakers, researchers, and technology managers about how they use or could use global technology assessments.

While this inquiry was motivated by NSF's specific desire to understand the viability of the current World Technology Evaluation Center (WTEC) mission in light of the changes in the global S&T landscape, the focus of this report includes the broader implications of the findings for all groups providing international S&T information (ISTI) services. Information specific to the inquiry on WTEC is provided in Appendices A and B. Finally, we reviewed and recommend metrics for assessing the outputs and outcomes of ISTI services.

BACKGROUND

S&T contributes to economic growth, human health, and a strong national security, as well as to the need of the human spirit to know about its surroundings and understand the universe. Scientific and technical knowledge flourish within an open atmosphere with information flowing quickly and easily to interested colleagues and researchers around the globe. Traditionally, scientific information has been freely shared through a formal publication process that results in a wide dissemination of information on the latest developments in specific areas of science.

Information about technology development, however, is not always easily obtained or widely available. Technical know-how is sometimes published in open literature but is more often embodied in products, in patents and trade secrets, or in the form of "tacit knowledge" in the heads of engineers and technicians working on a specific product. Sharing this information is just as important to progress as is sharing scientific information. Nevertheless, it takes more effort to identify, obtain, and transfer knowledge about technology, particularly if that knowledge is in a foreign language or located in a foreign laboratory.

Sharing S&T quickly and effectively can enhance the productivity of scientists and engineers. Information about research and development (R&D) on scientific or technical topics taking place in other countries can instruct researchers and managers on ways to maximize U.S. investments in research. Information about where foreign competitors are placing their bets can help a U.S. company either seek an alliance or better prepare to compete in the global marketplace. Because R&D is so important to the strength of the United States, the federal government has provided services to increase awareness of foreign S&T activities, as well as services to assess the state of foreign technology.

S&T information services are provided by a number of government and private organizations, including the Foreign Broadcast Information Service (FBIS) and the Office of Naval Research (ONR) technology information services, which collect both scientific and technical data. Several organizations focus principally on technology. One of these is WTEC,[1] at Loyola College in Baltimore, which provides assessments of foreign R&D in selected technologies under a cooperative agreement with the NSF, which funds and provides guidance to WTEC. Loyola's International Technology Research Institute (ITRI), with R. D. Shelton as director, is WTEC's umbrella organization.

---

[1] Formerly known as the Japanese Technology Evaluation Center (JTEC)

WTEC's mission includes informing U.S. policymakers, strategic planners, and industry technology officers about the status of selected technologies in foreign countries and how this compares to the status of technologies in the United States. WTEC assessments cover basic technology research, advanced development, and applications and commercialization.

From its inception in the 1980s, the WTEC program—and the other services that arose around the same time, i.e., the Stanford Research International's (SRI's) Washington Office and the Department of Commerce's (DoC's) Japanese Technical Literature Program (JTLP)—tended to focus more on the technology race between the United States and such countries as Japan and Germany. The perception in the 1980s that the United States was running in second or third place behind Germany in machine tool and medical equipment and behind Japan in manufacturing technologies contributed to the creation of WTEC and similar services. In the late 1990s, a different view prevails, that U.S technology is once again competitive with the best in the world.[2] Perhaps neither perception of technology weakness or strength is entirely accurate, but perceptions of U.S. competitiveness affect the demand for information about foreign S&T. It is safe to say that the demand for comparative technology from services like those WTEC provides is not taking place in the same context as it was when these services were first conceived. Given this shift, this is a good time to reassess the need and demand for global technology assessment, whether new customers are emerging, and how the needs of the traditional customer may have changed.

DEFINITIONS

Throughout this paper, for the ease of the reader, we often refer to all civilian ISTI collection and assessment activities simply as "ISTI." While we recognize that collection and assessment are distinctly different tasks—with collection services providing data and information and assessment services analyzing the data to create new knowledge—collection and assessment activities both provide information about foreign S&T. When the distinction between collection and assessment needs to be made, this is clearly stated in the text.

In addition, throughout the text, the terms R&D and S&T are used with these abbreviations. R&D refers to the practice of experimentation and the creation of new processes and products, while S&T refers to the body of knowledge about the physical world contained in the disciplines of science and in the application of engineering principles.

---

[2]This viewpoint is expanded upon in a RAND report: Popper, Wagner, and Larson (1998).

Finally, although neither *foreign* nor *international* is particularly fitting to describe the information we are discussing in this paper, we nonetheless use the terms somewhat interchangeably throughout the text. The term *global* may be a more appropriate to apply to the activities being examined, but this word has only recently come into usage with reference to S&T information. Accordingly, it is only applied to more recent activities.

ORGANIZATION OF THIS REPORT

In Chapter Two, we present the results of our literature review examining the historical efforts to collect ISTI and the transitions for the supply of, and demand for, ISTI over time. Chapter Three provides the results of our interviews on the needs of customers for global technology assessments. Chapter Four discusses options for responding to the changing circumstances and demands for global technology assessment revealed in the survey. Chapter Five discusses ways in which government services (such as WTEC) might account for their outputs and outcomes to comply with the Government Performance and Results Act of 1993 (GPRA).

Appendix A contains the results of the interview process directly applicable to WTEC. Appendix B contains a list of WTEC reports and tables of citations. Appendix C contains descriptions of scientific and technical assessment services from SRI, the Asian Technology Information Program (ATIP), the Microelectronics and Computer Technology Corporation (MCC), and ONR (Asia and Europe. Appendix D lists reports published by SRI, ATIP, MCC, and ONR.

Chapter Two

THE HISTORY OF PROVIDING ISTI SERVICES: RESPONDING TO CHANGING NEEDS

The U.S. government has been collecting and analyzing international civilian S&T since the Second World War. Over that time, the demand for ISTI has been affected by at least two transitions. In this chapter, we examine the two transitions in ISTI collection and assessment in detail and present the case showing that a third transition is now occurring.

THE ORIGINS OF GLOBAL TECHNOLOGY ASSESSMENT: THE TRANSITION

The U.S. government has funded programs to collect and assess foreign civilian S&T for over 30 years. The focus on civilian S&T emerged from similar functions performed in the U.S. defense and intelligence sectors during the Second World War. The earliest efforts to collect intelligence on ISTI were developed within British military intelligence by R. V. Jones.[1] Jones was able to use the openly available S&T information to assess the likelihood that Germany would develop specific weapons. His efforts proved decisive in determining several military threats just before and during the war. Although these first scientific intelligence efforts were initiated in Great Britain, the United States quickly

---

[1]R. V. Jones, working in British Intelligence in the later 1930s and 1940s, established the first scientific intelligence unit and established principles for the collection of information. His principles established the baseline against which subsequent services and activities would be compared:

> The primary problem of a Scientific Intelligence Service is to obtain early warning of the adoption of new weapons and methods by potential or actual enemies. To tackle this problem, it is advisable first to consider its nature. The adoption of a fundamentally new weapon proceeds through several stages: (1) General scientific research . . . (2) Someone in close touch with a Fighting Service . . . to think of an application of the results of academic research; (3) Ad hoc research and small scale trials are performed in a Service laboratory; (4) Large scale service trials are undertaken . . .(5) Adoption in Service. The first stage is generally public. . . . The third and fourth stages are more difficult to observe. . . . In many ways those are the most important stages at which a good Intelligence Service should be able to give timely warnings to its Government . . . . (Jones, 1978, p. 73.)

Among the many who considered Jones "the father of modern scientific and technical intelligence" was former Director of Central Intelligence, R. James Woolsey, quoted in the *Journal of Electronic Defense*, January 1998 edition, p. 30.

followed the British lead. After the war ended, scientific and technical intelligence was institutionalized in U.S. intelligence services.

One of the first scientific information services established in the United States was within the Office of Scientific Research and Development (OSRD). Dr. Vannevar Bush, a former Massachusetts Institute of Technology (MIT) vice president and dean of engineering who held several policymaking posts in Washington before accepting the lead at OSRD, directed that office during and just after the war. OSRD was highly effective during the war at funding R&D that led to products the military needed. In addition, OSRD monitored foreign S&T capabilities. OSRD research led to a number of scientific advances introduced during the war, including penicillin, radar, early calculating machines, jet engines, and atomic power. These developments proved so groundbreaking that, after the war, high-level policymakers turned their attention to applying S&T to address social problems and promote economic growth.

In a letter to Dr. Bush, President Franklin D. Roosevelt noted that OSRD "represents a unique experiment of teamwork and cooperation in coordinating scientific research and in applying existing scientific knowledge to the solution of the technical problems paramount in war." (Bush, 1960.) Roosevelt went on to suggest a postwar function using S&T to build a bridge between military and civilian applications. Prior to Roosevelt's recommendation, there had been negligible effort to transfer defense S&T information into civilian applications. Roosevelt's recommendation was a significant shift in government policy:

> There is . . . no reason why the lessons to be found in this experiment cannot be profitably employed in times of peace. The information, the techniques, and the research experience developed by the Office of Scientific Research and Development and by the thousands of scientists in the universities and in private industry, should be used in the days of peace ahead for the improvement of the national health, the creation of new enterprises bringing new jobs, and the betterment of the national standard of living. (Roosevelt, 1944.)

In his postwar treatise recommending a framework for a national science policy and the creation of a "National Research Foundation," Vannevar Bush built upon Roosevelt's recommendation by elaborating on government action to institutionalize R&D and information gathering. Among these actions, Bush called for government to encourage international S&T data exchange:

> International exchange of scientific information is of growing importance. Increasing specialization of science will make it more important than ever that scientists in this country keep continually ahead of developments abroad. In addition a flow of scientific information constitutes one facet of general international accord which should be cultivated.

> The Government can accomplish significant results in several ways: by aiding in the arrangement of international science congresses, in the official accredit-

ing of American scientists to such gatherings, in the official reception of foreign scientists of standing in this country, in making possible a rapid flow of technical information, including translation service, and possibly in the provision of international fellowships. Private foundations and other groups partially fulfill some of these functions at present, but their scope is incomplete and inadequate.

The Government should take an active role in promoting the international flow of scientific information. (Bush, 1960, Ch. 3, p. 22.)

As one of the functions of a "National Research Foundation"—a recommendation that eventually led to the creation of the NSF—Bush recommended that this entity "promote the dissemination of scientific and technical information and . . . further its international exchange." (Bush, 1960, p. 37, item m.) NSF's chartering legislation picks up the thread from the Bush report and authorizes the foundation to

provide a central clearinghouse for the collection, interpretation, and analysis of data on scientific and engineering resources and to provide a source of information for policy formulation by other agencies of the Federal government. . . . (U.S. Code Sec. 1862(a)(7).)

From the 1950s through the 1970s, government-sponsored efforts to track S&T worldwide were widely dispersed throughout the government.[2] Although the information collected on S&T was generally of good quality, it was often classified or buried within an agency that collected and used the information for mission-specific purposes.[3] As a result, the information had limited distribution, even within the U.S. government circles that needed this information.[4]

---

[2]During the 1950s and 1960s, tracking international S&T was conducted within the CIA in the Directorate for Science and Technology.

[3]Efforts to coordinate technical information can be traced back to the 1960s. In 1962, responding to a growing concern over management problems and diffusion of information created by the rapid growth of scientific and technical activities, the Federal Council on Science and Technology Policy established the Committee on Science and Technical Information (COSATI). COSATI's functions were to (1) coordinate agency scientific and technical information services; (2) examine interrelationships between existing information services, both inside and outside the government, and identify gaps or unnecessary overlaps; and (3) develop governmentwide standards and comparability among systems. COSATI members represented 12 of the largest scientific and technical information–producer and –user agencies of the government. COSATI encouraged voluntary coordination of U.S. scientific and technical activities abroad and collection of foreign technical information. According to a former COSATI chairman, sharing of foreign technology information among participating federal agencies was encouraged but not required, and distribution was presumed to be shared. In 1972, the White House Office of Scientific Research and Development transferred the leadership of COSATI to NSF as a management improvement; however, according to the U.S. General Accounting Office (GAO), there was a lack of support for the effort at NSF.

[4]Not infrequently, this information is so highly classified that only a few government officials with the proper clearances can read the reports.

MEETING THE JAPANESE CHALLENGE: THE SECOND TRANSITION

The second transition for ISTI came in the 1970s, when the "Japanese challenge" raised awareness of the increasingly competitive nature of technology-based products being manufactured abroad. The need and demand for information on civilian technologies led to the growth and expansion of government ISTI services in the civilian technologies.

In the late 1970s, two factors affected the demand for ISTI: (1) the increasingly competitive Japanese economy, and (2) the realization that national security is closely tied to the strength of the civilian economy. It is easy to forget now, at a time when U.S. companies have regained significant competitive strength in the global market (see Popper, Wagner, and Larsen, 1998, and Council on Competitiveness, 1998), how much alarm was raised about Japanese technological ascendancy from the 1970s through the early 1990s. By the mid-1970s, the United States had effectively lost its consumer electronics industry to Japan. In 1981, the United States was asking Japan to restrain auto exports. Other industries also came under increasing competitive pressures.

Some pundits shrugged these changes off as the natural cycle for "sunset" or "smokestack" industries, such as textiles and steel: The future was going to be in technology and services, they said. However, in the early 1980s, when the U.S. semiconductor industry looked to be in peril, the argument about sunset industries began to lose credibility, and demand for government action grew.

In a 1982 book, *Minding America's Business*, Robert Reich and Ira Magaziner wrote:

> Once dominant in most of the world's businesses, many U.S. companies have not kept pace in recent years with changes in the international competitive environment. Our systems for evaluating investment decisions have not sufficiently considered the competitive evolution of business. . . . We have allowed foreign competitors to gain advantages in other national markets from which they can better penetrate in U.S. markets. . . . By contrast, Japan and many European countries have adopted explicit policies for promoting selected businesses. (Reich and Magaziner, 1982, pp. 7–8.)

According to Reich and Magaziner, these policies included global information collection and targeted dissemination of that information to nationally based companies.

As the Japanese challenge unfolded, it became clear that Japanese and European businesses and governments were actively involved in collecting intelligence about U.S. technology. German and French firms have long had a tradition of gathering data on competitors' technologies and organizing this information into useful, strategic information that fed R&D and investment decisionmaking. Japan had an even longer tradition of seeking and importing foreign technology, dating at least from the Meiji era of the mid-19th century. In 1957, the government of Japan set up organizations, such as the Japan External Trade Organization (JETRO), as modern scientific information centers for Japanese business. By the early 1960s, all large Japanese multinationals had units collecting and assimilating foreign S&T information.

By the mid-1980s, JETRO, funded by the Japanese Ministry of International Trade and Industry (MITI), was sponsoring collection efforts aimed at 57 countries in 76 research offices around the world, targeting developments in 30 industries. Government funding for this effort totaled more than $180 million. (GAO, 1993, p. 5.) Several of the Japanese R&D consortia of the 1980s grew out of intelligence gathered from around the world about technology trends, markets, and Western vulnerabilities.

Despite the competitive challenge from abroad, the concept of tracking and monitoring civilian S&T in other countries still did not come easily to the United States. For some time after the Second World War, the United States was the only advanced nation with an intact science base. This meant that, for nearly three decades, the U.S. S&T base was by far the strongest in the world. The lag in world capability in science, technology, and engineering enabled the United States to develop world-class products. Even as other nations began to pull up next to the United States in science and engineering, a "not-invented-here" mentality seemed to dominate U.S. thinking. Decrying this tendency, Simon Ramo wrote in 1985 that

> All nations, with the exception of the United States, are convinced that effective use of technology, which is key to their future success, will depend upon how skillful they are at latching onto the technology that originates in other nations. Each nation knows that no one nation is big enough in terms of S&T to attain the preeminent position that the United States has enjoyed in the past. Consequently, they are building their policies around the basic concept of acquiring technology from outside their borders. Only the United States assumes, despite evidence that it is no longer true, that it has the technology and the other nations do not. . . . (Ramo, 1985, p. 14.)

During the mid-1980s, the notion that foreign technical capabilities not only posed a threat to the competitive posture of U.S. industry but also threatened to undermine national security emerged as a principle motivation for collecting ISTI. The idea began to emerge that U.S. power is tied, not just to military strength, but to economic strength as well and that the U.S. could not take its economic preeminence for granted. In 1988, Clyde Prestowitz wrote that

> [T]he United States has seemed to understand power and hegemony only in military and political terms, and has failed to note . . . that they can have economic dimensions as well. Today the real challenge to American power is not the sinister one from the Eastern bloc, but the friendly one from the Far East.

> U.S. industry is not withering in the face of Soviet competition and the Soviets are not sending shivers through our financial markets. Of course, the Soviet Union cannot be ignored—and we have not ignored them. . . . To the friendly challenge from the East, however, our response has been quite different.
> (Prestowitz, 1988, pp. 21–22.)

The concept that there are technologies—and supporting areas of science—that contribute disproportionately to a nation's competitive strength and national security led to a number of key actions by policymakers in the late 1980s. Congress responded in several ways. Among them was the Japan Technical Literature Act of 1986 (described in Appendix C) and the Omnibus Trade Act of 1988, both aimed at enhancing the competitiveness of key U.S. industries. In 1988, Congress also passed legislation requiring the compilation of reports on where the U.S. stood relative to the rest of the world in "critical" military and civilian technologies (42 U.S. Code Sec. 6683, 10 U.S. Code Sec. 2506 (b)(5)). These reports were intended to provide possible targets for U.S. government support.

With a growing awareness that "dual-use" technologies can be critical to national security and that U.S. security increasingly depended on the economic strength of key civilian industries, efforts to study foreign capabilities began to mature. The Department of Defense, which had supported most of the foreign technology assessment up until that time,[5] shifted its collection and analysis focus from purely defense technologies to any dual-use technologies that might have an impact on U.S. national security.

Moreover, as the Japanese challenge increased the demand for information and assessments of foreign science and/or technology, the U.S. government responded by establishing or expanding civilian technical information services and products for its own use, as well as for use by industry. This shift—which accelerated a change in focus from defense to dual-use and civilian technologies—was a significant transition point for U.S. ISTI services.

Nevertheless, the shift occurred slowly and resulted in the creation of several government and private-sector services, including the following:

- JTLP—created to monitor, acquire, and translate Japanese technical publications and make them available to the research and policy community

---

[5]In 1993, the GAO reported that

> There are 62 civilian and military offices and divisions within 6 departments and independent agencies that monitor foreign technology information. **U.S. efforts are primarily oriented towards the military** in terms of the number of organizations and the way resources are expended. (GAO, 1993, p. 5. Emphasis added.)

- ONR's Tokyo Office—created to provide a continuos feed to a team of researchers tracking foreign technical capabilities

- Office of Technology Assessment (OTA)—created to provide early indications of the probable beneficial and adverse impacts of the applications of technology.

But these services often lacked coordination. A 1992 report by the Carnegie Commission on Science, Technology, and Government joined the chorus of earlier reports and directives requesting government action to provide coordinated access to information about international S&T activities. In the chapter entitled "Functions: Field and Headquarters Activities," the Carnegie Commission recommended the following:

> The government should *monitor S&T developments abroad*, focusing on what the government itself needs to know. It also needs to help minimize barriers to the much more extensive monitoring and dissemination efforts undertaken directly by industry and academia. Indeed, the government must facilitate the national diffusion of open information from all sources. (Carnegie, 1992, p. 40. Emphasis in the original.)

The Commission further recommended that

> The government should *monitor and understand the S&T policies and strategies of other nations* and regional groupings. This may involve trade, research priorities, arms exports, or different assessment of the potential payoffs from the promotion of investments in various engineering fields. (Carnegie, 1992, p. 40. Emphasis in the original.)

A THIRD TRANSITION IN THE DEMAND FOR ISTI SERVICES

Collecting S&T information and assessing its meaning for U.S. government R&D spending and industry investments has evolved as the needs and demands of the security and scientific and engineering communities have shifted and changed. The brief history of the supply of and demand for ISTI reveals two significant transitions in the demand for ISTI, one occurring after the Second World War and the other occurring in the late 1970s and early 1980s, when services were established or redesigned in response to the Japanese challenge.

A brief examination of the current need for ISTI suggests we are in a third transition in the demand for ISTI and global technology assessment. The globalization of science and industry, the perceived renaissance of the technological strength of U.S. industry, and the impact of the information revolution are changing the landscape for ISTI services and how they will be delivered. We describe each of these factors below.

The Globalization of Industry

As world markets became increasingly interdependent in the 1980s and 1990s, U.S. government and industry recognized that the capacity of foreign countries to produce high-quality, globally competitive products, as well as their increasing ability to buy these products, called for a change in operating procedure. Seeking access to foreign technical capabilities and markets, U.S. industry accelerated the process of foreign direct investment—a phenomenon that has become equally robust coming into the United States as it is going to foreign countries. The resulting increase in trade and cross-investment has given rise to what some call the "globalization" of the world economy.

The global economy has grown in response to four factors: (1) the maturation of many national economies and the resulting market opportunities, (2) increasing technical excellence dispersed globally, (3) the facility of modern communications, and (4) the increasing complexity of modern technology. Accordingly, companies have found utility in seeking alliances with other firms to acquire, codevelop, or jointly market technology-based products. In addition, the trend toward foreign investment in R&D has grown considerably over the past decade. Finally, U.S. multinationals have built their own labs in foreign countries to be closer to markets and to access foreign talent.

Seeking Alliances. Companies are increasingly seeking global research partnerships as a way to strengthen core competencies and expand into technology fields considered critical for maintaining market share. Recent research shows that international R&D alliances are increasing sharply throughout the industrialized nations.[6] In the late 1980s and continuing into the 1990s, according to John Hagedorn, joint nonequity R&D agreements became the most important form of partnership. (NSB, 1998, p. 4-49.) The formation of these strategic technology partnerships has been particularly extensive among high-technology firms in such critical areas as information technologies, biotechnology, and new materials (Jankowski, 1998).

Investing in Foreign R&D. Firms invest in partnerships with foreign companies for many reasons, but as John Jankowski has noted, "all relate to growth in global innovation and the strategic need to establish networks for creating and strengthening firm-specific technological capabilities." (Jankowski, 1998.) As a group, foreign entities now invest significantly more in R&D than does the United States, a trend that is improving opportunities to access and use the results of foreign R&D. Jankowski found that, from 1987 to 1995, U.S. firms' investment in overseas R&D increased more than twice as fast as did company-

---

[6]Research by John Hagedorn, quoted in Jankowski (1998), p. 18.

funded R&D performed domestically (Jankowski, 1998, p. 19). While some of these investments improve access to markets for U.S. goods, others are strategic, seeking to tap into excellent research or manufacturing capabilities in foreign countries.

Building Multinational Labs around the World. A third part of the global phenomenon is the expansion of the conduct of international R&D by large multinational firms. According to the DoC, U.S.-based enterprises invest nearly $15 billion per year in offshore R&D, roughly 10 percent of their total R&D budgets. (Florida, 1998.) And according to Richard Florida, "Foreign-owned laboratories are a response in part to the rapid and thoroughgoing globalization of markets—in particular the fact that goods are increasingly produced where they are sold." (Florida, 1998, p. 31.)

International alliances, U.S. investment in foreign R&D, and the U.S. multinationals' investments in building foreign labs means that the large firms that can afford this mechanism of technology transfer have a great deal more direct access to foreign S&T know-how than ever before. On one hand, this may be lessening the demand for information from "eye witnesses" that U.S. government services have provided in the past. Opportunities for personal interaction increase the efficiency with which knowledge is transferred and mean that tacit knowledge—know-how that cannot be written down—can be more effectively shared. As Oliver Gassman and Maximilian von Zedtwitz have noted: "Informal know-how transfer by sharing experience between individuals is often superior to formal distribution of knowledge by means of hand-books, blueprints or patent rights." (Gassman and Zedtwitz, 1998, p. 131.)

U.S. Technological Strength

Along with the process of globalization, and perhaps because of it, perceptions of the relative strength of the U.S. technology base have changed considerably. As mentioned in Chapter One, the prevailing view among U.S. industry in the late 1990s is that U.S. technology is once again in a position of competitive strength.[7] However, statements about relative *national* strength no longer adequately represent the views of U.S. industry. Steven Popper, Caroline Wagner, and Eric Larson found that "given the global economy, [leadership in every technology area] would not be a significant concern for business or government...." (Popper, Wagner, and Larson, 1998, p. xviii.) U.S. industry no longer voices the notion that national S&T capabilities directly influence corporate competitiveness.

---

[7]This viewpoint is expanded upon in Popper, Wagner, and Larson, (1998).

The Information Revolution

The development of the Internet, electronic mail, and then the World Wide Web has drastically changed the way we view, desire, and receive information. Scientists, scholars, and engineers can directly exchange information with a touch of a key, allowing revolutionary ideas to flow through the digital highway. According to Nua Internet Surveys, world usage of the Internet is currently estimated to include 148 million people and more than 2.4 million Web sites, and its growth rate is expanding rapidly.[8] By mid-June 1998, Internet usage grew by 50 percent over 1997 growth levels and is expected to exceed 1997 growth rates of 84 percent.

Projected growth estimates show that, by the end of 2000, the Internet will have close to 350 million users. Though experts are hoping that electronic commerce will be one of the main uses of the Internet, users currently view the World Wide Web as primarily a source of information. The U.S. government has recognized the significance of this information revolution and its effect on ISTI information gathering and dissemination. In its 1993 report, *Technology for America's Economic Growth*, the White House committed itself to using information technology to promote the dissemination of federal scientific and technical data. Given the growth and use of information technology, perhaps it is time to take a closer look at how the Internet has changed the needs and access of customers in the dissemination of ISTI and how this change can be used to the advantage of ISTI providers and customers.

----

[8]Nua Internet Surveys, **http://www.nua.net/surveys** (last accessed September 1998).

Chapter Three

DETERMINING GLOBAL ASSESSMENT NEEDS
FOR THE FUTURE

Given the third transition in demand for ISTI described above, it is an important time to determine what users of ISTI information want now in terms of global assessments and what they will want in the future. In an effort to garner opinions and assess demand for ISTI, we asked a group of researchers and policymakers about how and why they use ISTI and what kinds of information they will need in the future. The individuals contacted were all users or were assumed to be users of WTEC services, since, as mentioned earlier, the NSF was particularly interested in how the changes in the needs for global assessment affected WTEC's current mission. Here, we report on the results of the questions that relate to broader issues of future needs and include but extend beyond WTEC services. While most of those responding to the issues of future needs use WTEC, they also use other services. The results of the WTEC part of the questionnaire are included in this document as Appendix A.

In this chapter, we discuss the interview methodology. In Chapter Four, we present options for responding to the survey results.

INTERVIEW METHODOLOGY

As a first step, we developed a list of possible contacts. The first source for contacts was a government personnel locator on the Web, where we identified a selection of heads of government R&D research units, as well as research scientists and government S&T analysts. We added names of people whom the team knew would be interested in ISTI. Two dozen names of members of the Society of Competitive Intelligence Professionals were also added to the list.[1] A number of WTEC workshop participants' names were included, as were the names of a number of sponsors. Several letters from sponsors were also used as input.

Next, we developed the following set of questions to use when contacting people for information:

_____

[1]SCIP is a private organization that promotes competitive intelligence.

- **Questions About WTEC/JTEC**

    1. Are you familiar with WTEC/JTEC? [If no, skip to question 8; if yes, continue.]

    2. Have you used any of their products? Which of WTEC's/JTEC's different products or activities have you taken advantage of?

    3. Are there particular studies that you recall as having been particularly good or influential? [If no, skip to question 6; if yes, continue.]

    4. Why was this report(s) influential?

    5. How did you use this report(s)?

    6. Can you give a specific example of the way a report was used to change policy or affect research funding?

    7. Overall, how would you judge WTEC's/JTEC's products?

- **Questions About Technology Assessment Generally**

    8. Are there other sources of global technology assessment that you use (e.g., MCC, ATIP, OTA, ONR)?

    9. How effective do you find the existing set of reports on global science and technology in meeting your needs for information about science and technology abroad?

    10. There has been a focus on Japan. Is there another geographic area that you feel should be better covered or emphasized?

    11. In your work, do you expect to continue to need information about global science and technology? If so, are there products or services that would better meet your needs than what is currently provided?

    12. Would private-sector sources meet your needs, and would you be willing to pay additional funds to get this kind of information?

    13. Do you have any other comments on the availability of global assessments of science and technology?

    14. Is there any one else you suggest we talk to about global science and technology assessment?

Note that this set of questions did not constitute a formal survey; the list of participants is not a scientific sampling. Therefore, the results are indicative but not statistically verifiable.

The team contacted 130 people in the course of seeking information on specific uses and needs for global technology assessment. All contacts were made by telephone. Of the 130 contacted, 91 agreed to participate. Of these, 52 were government policymakers, research scientists, or analysts and 39 were from the private or the nonprofit analytic sector. Two letters were also received, one from an industry source who had sponsored WTEC/JTEC studies while in government service, and one from an academic who had sponsored WTEC/JTEC studies while in industry. This brought the total number of participants to 93.

RESPONSES TO QUESTIONS ABOUT TECHNOLOGY ASSESSMENT GENERALLY (QUESTIONS 8–14)

Questions 8-14 were posed to participants in our attempt to characterize the types of technology assessments they used.

Other Sources of ISTI and Their Usefulness (Questions 8, 9, and 12)

Three of the questions asked about what sources of ISTI participants used. Based on the responses received, only one participant reported using WTEC/JTEC reports exclusively. Generally, participants reported using a range of sources to obtain information about global technology capabilities. Table 3.1 shows the distribution of how often the participants named a specific source. Some respondents mentioned more than one source.

When participants were asked if private-sector sources would meet all their needs and whether they would be willing to pay for these kinds of services, the responses were mixed. Among the 91 respondents, 70 had some comment on whether the private sector could meet most needs for this type of information, whether a mix between government and industry sources was best, or whether only government sources would meet the needs of their organization:

Table 3.1

Sources Used by Those Responding to Requests for Information

Information Source	Number Reporting Using This Source
Private-sector sources (e.g., Booz·Allen & Hamilton, Frost & Sullivan, Dataquest)	21
ONR reports	19
SRI S&T reporting (under government contract)	16
Academic journals, magazines, and newsletters	15
ATIP	12
Internet (e.g., articles on the World Wide Web and in on-line trade and press publications)	6
MCC services (member services only)	3
Other sources (e.g., conferences, FBIS, library research, interviews, travel, industry road maps)	20

- More respondents—27—thought that **a mix of government and private-sector** sources was more likely to meet their needs than private or government services alone. These respondents said that private sources are often more in-depth and strategic in approach but that government sources are less biased and provide a broad baseline.

- Almost the same number—26 respondents—said that **only government** sources would likely meet their needs. They cited the lack of bias in government and the comparative lower cost of government materials as reasons for their choice.

- Some respondents—16—felt strongly that **private-sector** sources are the most reliable and that they would tend to rely on these services most heavily or exclusively.

Of the 78 respondents who replied to the question about whether their current needs were being met by the mix of private, academic, and government reports available now, the largest number—40 respondents—said that their current needs are moderately well met by the current mix of information. Even so, many complained that detailed information is hard to find: Government overviews are good but are in need of more detail. A good number—25 respondents—said that their current needs are only fairly well met or are not being well met. In these cases, however, the respondents could often still name one or two services that still met their needs to some extent. Only four respondents to this question felt that their needs for ISTI were being well met.

Future Needs for ISTI Services (Questions 10, 11, and 13)

Overwhelmingly, the respondents reported that they will continue to need global technology assessment and foreign S&T information. Only four out of 91 respondents reported that they did not expect to continue to need ISTI, and, generally, these comments were made because the person was changing jobs or job responsibilities. Moreover, a number of those responding positively said that they expect their needs for this information to increase in the future.

When asked whether the future focus of ISTI should continue to expand beyond Japan to other parts of the world, the answer was also an overwhelming yes. Many respondents named specific countries or regions of the world beyond Japan that are of interest to them, as shown in Table 3.2. However, more often, respondents said that technology is global and that while the regional focus can sometimes be helpful, they more often need to know where the best R&D is being conducted around the world, regardless of the region or country. Respondents are generally more interested in knowing about technology developments than about capabilities in a specific country.

Table 3.2

Regions of Interest

Country or Region of Interest	Times Respondents Named This Country or Region as Important to Them
Europe (including former Soviet Union)	37
Asia (Taiwan, Korea, Malaysia, Singapore, and others)	22
China	14
South America	7
Russia	5
India	5
Australia	2
Africa	2

RESPONSES BY USER CATEGORY

Finally, to see if responses differed by the type of user, we grouped the individuals contacted for this study into five categories: (1) government research directors and policymakers who sponsored WTEC research, (2) government scientists and engineers, (3) government analysts, (4) private-sector workshop participants; and (5) private-sector potential users. Table 3.3 summarizes their responses to four key questions about technology assessment.

Table 3.3

Responses by User Category

Users	Types of Public and Private Sources Used	Degree to Which Current Sources Meet Information Needs
Government research directors and policymakers	Heavily rely on government sources	Good
Government scientists and engineers	Technical literature and public and private assessments	Poor
Government analysts	Full range of sources	Poor
Private-sector WTEC workshop participants	Full range of sources	Good
Private-sector potential WTEC users	Heavily rely on private-sector sources	Fair

With the exception of government scientists and engineers, all groups used a variety of sources for technology assessment information. Government scientists and engineers rely more heavily on technical literature, particularly on that available through the Internet. For example, one respondent noted that "[t]here is a paradigm shift with the Internet—more information is out there, so our need has diminished for assessments." Another respondent commented: "Excellent search engines have been developed to ferret out requested information . . . however, information on the Web still lacks the depth necessary for researchers like me to make informed decisions." In addition, more than half the government scientists and engineers interviewed say they use private services, with SRI being the source named most often.

Overall, government research directors and policymakers felt their current needs for global technology assessment are well met. While the two private-sector groups felt their needs were somewhat or usually met, government scientists and engineers and government analysts felt their needs were not met. Typical comments from the latter group include the following:

> Technology assessment is weak overall. SRI does a good job for the most part, but overall, reports are inexact and uneven—it's hard to get a cumulative picture. [There is] very little knowledge-building capability in the current information.

> By the time studies are published, it's too late—the technology is already ahead. They are not keeping up.

Finally, all the user groups felt that there was a strong need for the government to provide S&T information. Comments from the private-sector users reflect this belief. A respondent from the private-sector workshop group noted that "the programs provide enough information, and for a small company, it is good that government provides these workshops; otherwise, it would be too expensive." A respondent from the private-sector potential users group noted that "I can trust government statistics and analysis more than that coming from a private research house."

CONCLUSIONS

Based on the responses from users and potential users of ISTI, it appears that global technology assessment products are filling public needs and that there still is a role for government in supporting this service. Responses from several participants noted specific needs:

> Where would I go for information? It would depend on the type of information used; sometimes the private sources are not as believable as government sources.

Some of the private sources are biased; government is more objective . . . .

Government should be involved, but watch the bottom line—you can make studies forever.

I don't think that the private sector has either the motivation or the wherewithal to procure such specialized information.

What would it be like if this information were not around? For some people, it would be a real loss. . . . There is nothing to replace it.

Nevertheless, for most respondents, the need for information extended beyond a single country or region of the world:

The studies in the future should be technology based, not country based or focused. The central organizing focus should be technology—for example, what is the world biotech industry like at this time?

Best-in-the-world information would be useful. . . . Best-in-class is key to cooperating. Who the leaders are in different technologies would be good.

Tell me what is happening in technology around the world, not just in one place.

It is important to know the world status of certain technologies in order to feed strategic decisions.

However, existing government services, while judged useful, came in for criticism. Generally, criticism centered around three points: (1) government information is too broad or not timely enough; (2) the information is provided without adequate political or economic context; and (3) the information collection and dissemination are not well coordinated or easy to find.

Government information is useful, but once beyond the baseline, government information is probably only useful to government and not so much to the private sector.

I don't use government sources; they tend to be broad-based and not timely.

The information should venture further into the policy world—technical knowledge and policy knowledge are sometimes at odds. Existing information is not well related to policy themes that people in government struggle with.

What should government do? Government should continue to fund [ISTI], but agencies should coordinate across the government to share information and make it easily available.

Chapter Four

OPTIONS FOR RESPONDING TO USER NEEDS

As technology evolves and international research capabilities mature, the need for global technology assessment continues to increase. The demand for ISTI has grown to require broader-based information from around the world and about technology and the economic context within which it is found. Users continue to need information that is objective and unbiased, timely, and easily accessible. Emerging needs include assessments that look beyond traditional competitors to emerging capabilities worldwide and to best-in-class research anywhere in the world. Users continue to request better coordination among government ISTI sources, to access comprehensive information on a topic in a way that presents the information at varying levels of abstraction and detail. Most now see the Internet as the best way to disseminate this type of information.

NEW CHALLENGES FOR GOVERNMENT IN PROVIDING ISTI NEEDS

The government should play an important role in collecting, disseminating, and analyzing ISTI. This is the consensus of government and industry reports (both historical and current) and respondents contacted for this study. Nevertheless, most sources criticize the way in which these services have been managed in the past. Aside from pockets of excellence, such as WTEC, the efforts to coordinate ISTI functions have never been done well, and the information collection has been weakly coordinated. Structural and organizational barriers have limited government's capacity to provide this service.

Admittedly, collecting, analyzing, and disseminating information on all areas of S&T worldwide are huge and complex tasks. Not only is there a broad geographic area to cover, but the needs of users vary considerably across the government and between the government and the private sector. This is further complicated in that breakthroughs in a number of key areas of S&T develop so quickly, it is difficult to keep up with them.

Deciding where to focus collection efforts, both technically and geographically is also challenging. Different areas of S&T require different approaches to understanding, analyzing, and disseminating information. For example, science can be tracked using bibliographic materials, but technology often requires personal contact—a method of information collection that WTEC and MCC have both used effectively. While face-to-face meetings and laboratory visits appear to be effective, this is an expensive way to gather information.

While these challenges are not receding, current conditions present an opportunity for facilitating collection, analysis, and dissemination of ISTI. A confluence of factors is changing the landscape for ISTI: Information is more likely to be unclassified than it has been in the past; more opportunities exist to tap knowledge being gathered by global travelers and denizens than ever before; information technology is finally at a stage where true coordination could be made to happen; and new knowledge-management tools are designed to handle just this type of information challenge. The next section discusses options for using new opportunities to improve ISTI.

In examining existing services, it would appear that the WTEC mission—providing information on the status of selected technologies in foreign countries—could be accomplished in several ways. These include using the WTEC model and relying principally on government, relying on a combination of government and industry, and relying on industry sources only. A number of private-sector groups provide similar services for a fee. A majority of respondents, however, thought that a mix of government and private-sector sources was more likely to meet their needs than either private or government services alone.

Despite existing competition among services, it is possible to build more competition within the WTEC service itself. One way to do this would be for WTEC managers to seek alliances with similar services, such as MCC or SRI. For each assessment opportunity that comes up, a "bidding process" could take place within the alliance for the group that can conduct and disseminate the assessment most efficiently. This would help build networked relationships among these providers, possibly serving their customers better. It may also take advantage of varying technical strengths among the groups. Finally, it may help bring costs down as each group attempts to offer more-efficient services to win business.

It is also possible to consider alternative models that would use new information-management tools and personal and telecommunications networking to make the ISTI function more efficiently. One improvement would be to use Internet search capabilities to conduct "smart" searches of bibliographic information for the individual user. Another would be applying software systems to query various databases and provide coordinated responses without having to build a new database. A third possibility would be to use information-management tools to hold "virtual meetings" of experts—a method of seeking expert judgment that could be used on either a periodic or a continuous basis. Finally, parsing information in a "tiered system" of information presentation, from less to more technical, would meet the needs of a broader range of users.

GOVERNMENT OPTIONS FOR MEETING THE NEEDS OF THE THIRD-TRANSITION USERS

Government options for meeting the needs of third-transition users of ISTI range across the possibilities offered by new knowledge-management tools. Both technical literature and expert judgment are needed to evaluate ISTI. This is a common feature of all technology assessment efforts. These methods are still required to conduct a true technology assessment. No matter how smart, computers will not be able to supplant expert judgment about technical capabilities.

This does not mean, however, that the function of conducting assessment cannot be made more efficient. After all, modern banks, as an example, fulfill their functions more efficiently, streamlining the time and effort that the customer has to put into depositing checks, paying bills, and getting cash. The principal function—managing money—is the same, but the method of service delivery is considerably more efficient. The same types of tools are available to the government to apply to the management of S&T information.

Applying New Tools: Information Gathering and Coordination

The first step in conducting global technology assessment is surveying literature. Technical literature is currently available in libraries, in government offices, and in analytic institutions. Increasingly, however, this information (or at least bibliographic data) is available on line. The challenge here is to use existing software and knowledge-management techniques to create a "smart" search that will more quickly and easily identify information related to the subject of interest. For example, the National Library of Medicine has a service called "Medline," which releases onto the World Wide Web a "knowbot" that searches for up to 200 articles or papers related to the subject in question. The knowbot then "sees" what, of that search, is of greatest interest to the user and goes out again, with a refined search, to find additional material. Similarly, the on-line edition of *Science Magazine* has a box on its search page that says "Alert me when new articles appear on this topic." This kind of service would not only be valuable to WTEC panels and staff seeking information, it would also be of great value to many of the users with whom we spoke.

The GAO (1990) noted that one obstacle to coordinating S&T information services had been that varying software protocols are used to store and retrieve information. While this is still true today, new metatools are being developed that can supersede existing software protocols. It is possible to buy or build a platform that can search data in whatever form it is stored. Moreover, using new tools, it is not necessary to build a new database. A universal search capacity could be developed that accesses any government and private service. Using tools like this, it may be possible to tap the different sources of ISTI for information relevant to a specific query, creating a "virtual" coordination service that creates no new programming.

One of the obstacles to coordination is the fee-for-service status of a number of the ISTI services on the World Wide Web. Even the government National Technical Information Service's (NTIS's) on-line site requires a subscription fee and password. This constitutes a challenge to virtual coordination, but one that can be overcome. Two options are (1) having the government ISTI coordination search capability acquire a blanket subscription to these different government services so that those using the service would be able to access information or (2) having the government ISTI service bring users to the "front door" of a pay service.

Applying New Tools: Seeking Expert Judgment

Information technology is suitable for the transfer and storage of explicit knowledge. However, little is known about how information technology can be used for creating and sharing tacit knowledge. This is the kind of knowledge embedded in the heads of engineers and other experts—information that is difficult to tap but that is crucial to understanding some key technologies. No computer search capability can transmit this tacit knowledge.

Applying information tools and knowledge-management principles to seeking expert knowledge and judgment is more challenging than developing the "virtual ISTI coordination" described above. While printed materials will continue to be valuable, the importance of interpersonal contacts as a method of conveying knowledge may increase as technology development accelerates. Information experts have noted that the shift to computers as a way of delivering information and knowledge resources has already changed the way in which we create, record, supply, and use knowledge resources. We are moving away from the traditional publishing model, in which large amounts of information are supplied to individuals who must then extract knowledge from that information at great effort, toward a more "organic" model that reflects an interactive relationship between the knowledge-holder and the knowledge-seeker.

One of the characteristics of knowledge management emerging out of the information revolution is the need to combine formal and informal ("expert") knowledge. Doing so will require some significant changes in how knowledge-haves and knowledge-seekers use the computing environment to reap the full potential benefits of developing and sharing information to create new knowledge. The environment will have to become truly "knowledge centered" instead of "document centric." This means tapping expert judgment.

The challenge to technology assessment is significant, but the potential rewards of using new communication tools like this are also great. Consider an on-line "hyperforum" that allows experts to meet in "virtual" space to discuss technical developments in a foreign country: Expert residents in that country could be brought on line for a brief conference to share information. If face-to-face communication is needed, a videoconference can be effective, but usually only after the participants have met in person. "Virtual meetings" could become a tool that is used on a periodic and continuous basis, which would be better suited than "snapshots" to the ongoing nature of technology development and transfer.

Applying New Tools: Dissemination

Using Web sites to disseminate documents is the most well-used feature of the Web at this time. There is a great deal of material on the Web that can be found, with some effort, by a persistent searcher or by someone who knows where to find it. Nevertheless, as Web-based dissemination evolves, the Web will no longer be used as a place to post an electronic version of a published book. Instead, information will be "tiered" from a level of abstraction down through levels of increasing detail. As this feature matures, it will find particular application for ISTI. Users could move down a "tiered system" to the level of detail that they need. An abstract would be the first "tier," followed by increasing levels of textual and graphical detail. For a policy audience, pointers to documents providing political and economic "context" could be added. The more technical user would find pointers to other sites with additional information or a bibliography.

SUMMARY OF RECOMMENDATIONS

Interviews with both industry and WTEC users and RAND's literature analysis have resulted in several overall recommendations toward the government's role in assessing technology:

1. Continue government involvement in global technology assessment and accompanying services that collect ISTI.

2. Continue NSF involvement in supporting global technology assessment, but expand formal planning and administration to include the DoC.

3. Consider creating a government-wide coalition of agencies conducting technology assessment and S&T collection or using ISTI; task the coalition to explore the creation of a virtual network for cooperative exchange of timely, relevant global S&T information.

4. Through a government-wide coalition of agencies working with WTEC, consider the utility and technical possibility of creating a government-wide S&T information service center.

Chapter Five

ASSESSING THE OUTPUTS AND OUTCOMES OF GLOBAL TECHNOLOGY ASSESSMENT

Third-transition ISTI services will require new ways of measuring outputs and outcomes. This is particularly important in the new era of accountability introduced with the 1993 passage of GPRA.

A NEW ERA OF ACCOUNTABILITY

U.S. government policymakers and program directors are increasingly asked to account for the outputs and outcomes of programmatic activities. Since 1993, bipartisan efforts spanning both the Legislative and the Executive branches have resulted in a set of requirements for accountability in government. The new requirements are embodied in: (1) the Clinton administration's National Performance Review; (2) GPRA and other legislation; (3) recent congressional committee requests for reports; and (4) policy documents, such as *Science in the National Interest* (White House, 1994). Government programs are required to develop a statement of goals and to identify and apply measures that will represent both the outputs (products) and the outcomes (impact) of government activities. The principal idea is that the continued feedback the measures provide will help program directors manage their activities better and more efficiently.

Methods to assess the outputs of scientific activity have been receiving increasing attention in literature over the past decade as data collection has improved and as analysts have refined direct and indirect measures and indicators. Most of the output and impact measures we examined for a previous study are based on some form of bibliometric analysis (Wagner, 1995). Bibliometrics can find a place in assessing global technology assessment, but other measures are also needed.

Applying Measures to Assessment Activities

Global technology assessment shares some features with basic research: For both activities, the goal is increased knowledge and understanding. Like research, this goal is multidimensional and not easily amenable to direct measurement. But, like research, global technology assessment can be measured indirectly using customized qualitative and quantitative metrics that examine both short term, product-oriented information and longer-term impacts.

The principal features of knowledge creation subject to evaluation are the (1) efficiency of information gathering and production, (2) quality of the verbal and written reports, (3) timeliness of information delivery, (4) accessibility of information, and (5) impact of information. Each of these factors can be measured using qualitative and quantitative assessment tools. The text below presents useful measures. The conclusion discusses how these different measures could be applied to the global technology-assessment services.

Qualitative Measures

Qualitative measures for assessing the creation of new knowledge have been heavily, indeed almost exclusively, used to assess knowledge-generating activities. These methods include peer review, expert judgment, and user feedback. Each of these is described briefly below.

Peer Review. Peer review is the most fully developed of the different qualitative measures available for research and knowledge-creation activities. It has been used effectively to assess WTEC program funding in the past. Peer review is generally an ad hoc activity, used most often when choosing to sponsor or to renew an activity. Writing in *Science Magazine*, Ron Kostoff (1997) has recommended that peer review be used as the dominant metric applied to basic research and knowledge-creation activities under the new regime created by GPRA. Kostoff has developed a series of principles that should be used when developing a peer review for research and knowledge-creation activities. The three main criteria for determining the merit of a knowledge-creation activity are (1) project merit, (2) richness of the approach, and (3) team quality. Other principles that Kostoff suggests should be applied when structuring a peer review or expert judgment committee include the following (Kostoff, 1997):

1. acquiring the commitment of the reviewing organization's senior management to sponsoring a high-quality review (this can mean both financial commitment and action creating incentives for staff to meet specific criteria)

2. ensuring that the review manager and the proposing team are both motivated to conduct or participate in a technically credible peer review

3. choosing the peer reviewers based on competence and objectivity

4. determining up front the relevance of the activity being reviewed to the mission of the sponsoring organization

5. requiring reviewer anonymity and reviewee nonanonymity

6. maintaining a balance between the cost of the review and the quality of the reviewing team.

Peer review has been applied to past WTEC activities to determine quality and relevance. Peer review is also the main methodology WTEC uses to assess the quality of international S&T. Nevertheless, peer review has not been used on all global technology collection and assessment activities that the U.S. government has sponsored. (For a description of some of these activities, see Appendix D.) Future government support for global technology assessment would benefit from a standardized set of peer review principles when choosing a team and funding global technology assessment.

Peer review has several drawbacks with relation to better program accountability. First, peer review is often conducted before a project or program starts. Thus, unless the review is evaluating *additional* funding for an existing project, findings of a peer review panel are not always relevant to reporting outputs and outcomes, per GPRA requirements. Secondly, peer review, by definition, taps expert input; usually, the interests of stakeholders are not represented on peer review panels. GPRA, in part, seeks to ensure that government activities have broad relevance—ad hoc peer review does not allow stakeholders to assess the value of publicly funded activities.

Expert Judgment. Expert judgment is a form of peer review applied to activities where experts are either convened or surveyed to assess midcourse progress or *post hoc* quality of a publicly funded research activity. Expert judgment, although used less often than peer review, focuses on the outputs and outcomes of knowledge-generating activities. An expert judgment team should be created using principles similar to the ones above, but this team would also consider data on actual project activities. Expert judgment is perhaps more directly responsive to the GPRA requirements than peer review, which generally does not focus on outputs and outcomes. However, it is quite expensive to convene expert judges. Using methods to routinize data collection (discussed above) and on-line consultations combined with a schedule for asking experts to examine activities, may help reduce the costs of these activities.

User Surveys. User surveys have been applied to a whole range of service-oriented activities. The WTEC program has used them effectively for example, to assess the quality and impact of services and products. User surveys can provide highly useful short-term feedback to program managers. This information can help management and sponsors make midcourse correction and program revisions. Generally, however, only interested parties submit user surveys. Accordingly, unless an independent third party conducts the survey and includes a broad group of respondents, the information is not directly relevant to the requirements of GPRA. Furthermore, the Paperwork Reduction Act of 1995 has made it more difficult for government departments to use surveys and questionnaires in a timely and effective manner. It often takes months to receive approval from the Office of Management and Budget for a proposed study.[1]

Quantitative Measures

Quantitative metrics are less well-developed—and much more controversial among researchers and program managers—than are qualitative measures. Admittedly, quantitative measures can be taken at face value and can sometimes be used to discredit a publicly funded activity. Nevertheless, when properly scoped and collected and when placed within context, quantitative measures can be more compelling and more politically sellable than qualitative assessment. In addition, both skeptics and supporters can reproduce the methodology for creating quantitative measures and thus can use it from year to year to track and monitor progress over time.

In the short term, input and output data representing the production of new knowledge creation can be measured. Inputs can be expressed in terms of funding and staff time, as well as in terms of research conducted, inquiries made, and visits sponsored. Input measures provide only limited information but, for purposes of measurement, can be compared to previous years' activities or like activities of similar services. To provide a fuller picture, input measures are joined with output measures.

Output measures include the various expressions and embodiments of new knowledge, such as numbers of papers, changes in research funding based upon new knowledge, or speeches given using the new information. The simplest metric for new knowledge creation would quantify the total number of reports created against the amount of funds provided to create the reports. While this measure provides some actionable information, this simple metric gives no indication of the efficiency of the process. Either an absolute measure is needed (the number of reports produced measured against the number of reports expected to be produced), or a relative measure could be applied (the level of efficiency of production compared to similar services).

The simplest input-output metrics are collected at or near the time an activity is performed and can be quickly reported. For example, if 10 reports are expected to be produced for $1 million in 12 months, this information can be submitted to the sponsor at the end of the period. A midpoint check can also show

---

[1]Comment from Phyllis Genther Yoshida, Director of the Asia Pacific Technology Program at the DoC.

progress toward the goal. Other measures include the extent to which new knowledge-based products are reported in trade journals or cited in technical literature.

Numbers representing the production of reports, as with most quantitative metrics, must be married to qualitative information to provide a full picture. Direct, short-term metrics, while important in describing the efficiency of outputs, cannot provide the whole picture. They must be placed in context and be compared against both the absolute goals (number of reports expected) and relative goals (produce more reports than other services). If, for example, the provider falls short of its goal of 10 reports in 12 months, this may be due to a lack of efficiency on one hand or may be due to the fact that one of the topics required a more in-depth treatment. The quality of the material clearly provides an important checkpoint in understanding the quantitative data. The lesson here is that measures must have both a "magnitude" component and a "quality" component.

To be truly useful to the sponsor, input-output measures should be combined with an efficiency plan. An efficiency plan would help determine how much improvement potential a service provider has. At the point of setting goals for assessment services, a benchmark study would be useful in determining whether a global technology assessment service is producing its reports efficiently. An initial overview of the WTEC program, for example, would appear to indicate that this service has been highly efficient in producing a product with available funds. Nevertheless, new information technology and other tools could introduce efficiencies into the process. A benchmark study may turn up goals for improved services.

Impact Measures

Clearly, the primary interest of government when funding global technology assessment is in the longer-term impacts and influence resulting from the information created. These impacts include successful outcomes of research, the judicious investment of government research funds, the strength of the U.S. S&T base, and the rate of technological innovation. Understanding and reporting the impact of ISTI on these larger national goals would go a long way toward providing the rationale for funding global technology assessment.

Complicating this type of measure, however, is the fact that the longer-term impacts and outcomes of knowledge creation are influenced by a multitude of factors that are nearly impossible to account for: other sources of information, advances in science that make a published paper irrelevant, financial conditions, serendipitous developments, synergies emerging from collaborations, and so on. These exogenous factors make it exceedingly difficult to disentangle the impact of a government-sponsored report or program on a range of activity as broad as the national science base.

Focusing impact assessment on a narrower range of activity may still provide impact measures. These quantitative impact measures can include cost-benefit analyses or quantifying the economic benefit of user access to new knowledge. These types of assessments can help justify government spending but are not likely to produce information that is useful to sponsors or program managers, particularly given the costs involved in conducting a study such as this. Qualitative impact measures can include long-term assessment activities, such as the NSF TRACES study or other analysis that tracks the impact of a report on scientific development over a period of time. Qualitative impact measures can also include anecdotal information about the impact of a report on activities of a specific government or private-sector laboratory.

Ron Kostoff (1997) has found that the managers who use metrics have very eclectic requirements. They need "suites of metrics," often combined with other evaluation approaches, to perform comprehensive, multifaceted evaluations. Sponsors and program managers customize most of these sets of metrics to represent the activities of their specific programs.

Matching Measurement Tools to Goals

Both qualitative and quantitative measures can be applied to goals established for improved assessment services. Assuming that global technology-assessment activities can improve services in the five ways listed above (efficiency, quality, timeliness, accessibility, and impact), this subsection discusses measures that may be appropriate to meet these goals.

Efficiency of Information Gathering and Production. The importance of establishing a benchmark is most pointed when discussing an efficiency improvement: Perhaps the program being assessed is already operating at its most efficient level. Nevertheless, if it is determined that efficiencies can be gained, several *quantitative* measures would likely be most effective in reporting the outcomes of efficiency improvement. Establishing a goal, such as increasing the number of input sources or increasing the number of short reports by one-quarter over the previous year without additional funds can provide the blueprint for an eventual measure. The measure would be a simple input-output measure showing improvement in output (more information or reports) at the same level of input (staff or funds).

Quality of Oral and Written Reports. Quality assessments of new knowledge creation are difficult to make and tend to be subjective. It is often difficult to assess the objectivity of the respondents who are asked to judge the quality of an oral or written report. Nevertheless, quality measures are among those any group most prizes and would be particularly useful to those producing new information. Quality can be measured using both quantitative and qualitative measures. Quantitative measures could include bibliometric data, such as numbers of citations in the trade press, journals, and speeches. The usefulness of bibliometric measures for global technology-assessment products would need to be carefully considered, since much of technology assessment serves as input to active decisionmaking rather than to academic studies. Even so, the extent to which a study is positively cited in the trade press is an indicator of quality. The more useful information on quality would most likely be gained from expert judgment and survey data. WTEC and many other services already use qualitative measures like these to provide indicators of quality. Periodic use of an independent and objective survey tool may add to the validity of the information.

Timeliness of Information. Providing information to users more quickly than in the past or than other services do would be relatively easy to measure. Setting a goal, such as using knowledge-management tools to gather background information more quickly—say in four instead of eight weeks—and providing written material more quickly—say within one instead of three months—are goals that can be tracked and monitored over time. Reducing the amount of time required to issue a written report would also be a goal amenable to quantitative measures.

Accessibility of Information. As knowledge- and information-management tools improve, the ability to improve customer access to information greatly increases. WTEC, for example, reports that its World Wide Web page received 5,000 more hits in September 1998 than in August 1998. Using the World Wide Web to improve the accessibility of information is one possible goal that can be readily measured. Another goal would be to enhance the usability of information by creating abstracts, summaries, and data tables. Establishing a "tiered" information system for information (both old and new) could also be represented quantitatively. As a by-product of this method, this system could also produce statistics on the number of hits a given report receives. Thus the method could record how often users were willing to pay a premium for a given report. A qualitative report on the acceptance of this service among users would also be a good measure.

Breadth and Depth of the Impact of Information. As we have noted, the *impact* of global technology assessment services is among the most valuable, and yet the most difficult, information to generate. A qualitative measure of impact might be a market test for new reports. Asking users to pay for reports while the information is most timely, say within the first three weeks or months, would indicate the value of the information to users. Another quantitative market test, already in use by both WTEC and MCC, is the willingness of sponsors to support a research effort. Qualitative measures, such as testimonials and user surveys, are generally of use to program managers but are not useful as a metric for external justification. Testimonials can appear self-serving and cannot easily be placed in a broader context.

Setting Goals and Measuring Progress for an Alternative Approach

Should the vision for an alternative "smart system" of collection, analysis, and dissemination of ISTI, described above, be instituted, setting goals and tracking progress would actually be relatively less complicated than developing measures for existing projects. Implementing a new system like the one above could set goals and measures, such as those illustrated in Table 5.1. An automated "smart" system could collect these measures on a continual basis, so that reporting becomes a matter of printing out a brief report.

Table 5.1

Sample Goals and Measures

Goals	Measures
Create a cross-government coordination system for ISTI that accesses at least five government information services on line.	Demonstrate software that crosses existing data sets and accesses disparate data.
Reduce the access time it takes for a user to access a primary report and four related reports.	After setting a baseline for current average access time, demonstrate that new tools created for a service like this significantly reduce time of access.
Increase the number of users having access to government ISTI services by 20 percent in one year.	After setting a baseline as to how many users buy reports or access on-line services, count the number of visitors to the new service.
Using new tools, increase the interaction between government and contractor staff and experts in technical fields by 30 percent over one year.	Using previous WTEC panels as a baseline, count how many more experts have been contacted for their opinion than in the past.
Expand coverage of technical topics to cover the top three to five centers of excellence anywhere in the world.	Show reports that now cover important developments in more than one country or region.

RESULTS OF RESPONSE TO WTEC QUESTIONS ON STUDY USER INTERVIEWS

Among the different ISTI assessment services that grew up in response to the Japanese challenge, the most in-depth and highest quality material has come from WTEC. WTEC's mission is "to inform U.S. policymakers, strategic planners, and industry technology officers about selected technologies in foreign countries in comparison to the United States."[1] WTEC assessments cover basic research, advanced development, and applications and commercialization activities in different countries.

One of the objectives of this report has been to understand the viability of the current WTEC mission in light of the changes in demand for global technology assessments. To this end, the project identified WTEC customers and conducted interviews. Following a discussion of WTEC's history and its operations, we present an analysis of the impact of WTEC reports based on a literature analysis and the results of the interviews.

HISTORY OF THE JTEC AND WTEC

In the wake of the Japanese challenge, as policymakers considered options for reorganizing the U.S. system of scientific and technical cooperation and information dissemination (a project that would take more than a decade to implement), key congressional leaders felt that actions needed to be taken before the United States lost its competitive edge. Through the action of several federal agencies, the Japanese Technology Evaluation Program (JTECH) was established in 1983 within the DoC. Its mission was to gather and disseminate information about foreign technology. During its two-year operation in the DoC, the program staff realized that the Japanese were willing to establish a two-way exchange of scientific and technical information between the two nations.

---

[1]WTEC Web pages: **http://itri.loyola.edu** (last accessed March 1999).

In 1985, when the NSF assumed leadership of JTECH, its mission included promoting the mutual exchange of scientific and technical information between the United States and Japan. However, the NSF, realizing that the program needed some new direction and initiative, placed the program under a competitive grant. In 1989, the ITRI at Loyola College assumed management of the program and established the JTEC.

ITRI made several important changes to JTEC. Beginning in 1990, the program began to broaden its focus to include other countries. At that time, a new program, WTEC, was created. The new center reflected the changing global environment by focusing on Europe, the former Soviet Union, and Canada.

Both programs are now subsumed under the single name, WTEC. The ITRI staff at Loyola College help select topics, recruit expert panelists, arrange study visits to foreign laboratories, organize workshop presentations, and edit and disseminate the final reports. ITRI also broadened dissemination of the JTEC series and later the WTEC series of reports. Most recently, ITRI has created a World Wide Web site and CD-ROMS to disseminate full-text reports.

In recent years, ITRI has also begun soliciting private partners under its WTEC Community-Initiated State-of-the-Art Reviews (CISAR) program. Under this program, the research community proposes topics. If a joint NSF-ITRI selection committee deems the project worthy, the CISAR partner defines the study, obtains industry support, chooses a panel, and drafts a report. ITRI provides financial support, management, editing, and dissemination of material. WTEC (1997) is an example of a CISAR study.

Since ITRI has managed the WTEC program, 50 reports have been released on a wide range of topics. In the earliest days of JTECH, the program sought proposals on any technology area that was of concern to the group proposing the work.[2] Later, the government staff working on JTECH and JTEC began to identify ahead of time technologies that would be important to examine. In the early days, a team of government researchers and program staff determined the best proposals for funding. Later, a government sponsor or a group of sponsors proposed a project, and ITRI staff determined the merits of the proposal.

WTEC OPERATIONS

Here, we briefly examine how WTEC selects and conducts projects. Under the JTECH program, assessments were based on panelists' evaluation of current literature. However, once ITRI took over, a new method was introduced. WTEC

---

[2]Information gathered from an October 18, 1998, conversation with Tom Kusuda, former JTECH and JTLP staff member.

began to provide a "snapshot in time" by inspecting technology being developed in foreign laboratories.

WTEC assessments are done by small panels of about six technical experts who are leading authorities in their field, technically active, and knowledgeable about U.S. and foreign research programs. As part of the assessment process, panels visit and carry out extensive discussions with foreign scientists and engineers in universities and in industry and government laboratories. The visits are usually between one to two weeks long, with the panelists being divided into two or more subgroups. Each subgroup usually visits two sites per day—covering as many as 20 sites in one week.

Once the panel concludes its visit, a first-draft site report is circulated among the panelists and then sent to the site hosts for a review of their own institution. During this process, hosts are invited to correct errors and to request deletion of any sensitive material from the draft site report before it is circulated. Once the first revisions are made, the panel presents its preliminary findings at a one-day workshop in the Washington, DC area. Visual aids are distributed in hard copy at the workshop and are posted on the Web within a couple of weeks. The panel then publishes a report about a year and a half from the time of the workshop.

WTEC CUSTOMERS AND SPONSORS

WTEC's sponsors are mostly in government agencies. Its customers include individuals in government, industry, and academia.

Government Sponsors

In addition to the NSF, several other government agencies and departments sponsor WTEC studies. Past sponsors have included the DoC, the Defense Advanced Research Projects Agency (DARPA), ONR, the Department of Energy (DoE), the National Aeronautics and Space Administration (NASA), Air Force Systems Command (AFSC), the Air Force Office of Scientific Research (OSR), the Army Research Office (ARO), the State Department, the National Institutes of Health (NIH), and the National Institute of Standards and Technology (NIST). Table A.1 lists agencies and the number of reports they sponsored. Industry sponsors include the National Electronics Manufacturing Initiatives and Society of Manufacturing Engineers.

ANALYSIS OF THE IMPACT OF WTEC REPORTS

By early 1998, under the direction of ITRI, WTEC had completed 36 studies in materials, manufacturing, construction, space, sea, and information technologies. These reports focused on the areas described in Table A.2.

Table A.1

Government Agencies and the WTEC
Reports They Sponsored (1984–1998)

Agency	Reports Cosponsored
DARPA and OSD	24
DoC	20
DoE	12
NASA	6
ONR	7
NIST	3
AFSC and ORS	3
ARO	2
NIH	1
Department of State	1

NOTE: NSF cosponsors all WTEC studies.

Table A.2

Regional Concentrations of WTEC Reports

Focus Area	Reports
Japan	25
Europe	1
Europe and Japan	2
Japan and Germany	1
Canada	1
Mix (Western Europe, Russia, Japan)	3
The Pacific Rim	1
Russia and Eastern Europe	2

Of these reports, three stand out as the most popular and influential based on reference searches: *Satellite Communications* (1993); *Optoelectronics in Japan and the United States* (1996); and *R&D Status and Trends in Nanoparticles, Nanostructured Materials, and Nanodevices in the U.S.* (1997), all from ITRI.

The plethora of reports available from WTEC has prevented the full mention of each report; however, Appendix B of this study lists the reports from the JTECH and WTEC/JTEC program.

In *Satellite Communications*, WTEC surveyed the development of advanced satellite-communication technologies in Europe, Japan, and Russia. Specifically, the study focused on the experimental and advanced technology being developed in R&D and demonstration programs for commercial use. The panel concluded that the United States had lost the leading edge in the development of critical satellite-communication technologies. Furthermore, the study showed that, unless the United States upgraded its technology, it would also lose its leading share of the market.

Optoelectronics in Japan and the United States provided a comparison and review of Japanese and U.S. optoelectronics technology and manufacturing. Specifically, the report covered optoelectronic systems (particularly telecommunications, local-area networks, and optical interconnections), optical storage technology, waveguide devices and packaging, photonic devices and materials, optical sensor technology and specialty fibers, and economic considerations. The panel found that, while Japan had the lead in consumer optoelectronics, the United States had the lead in custom electronics. However, the study also found that the Japanese government had a clearer understanding of the role of optoelectronics technology in the development of future electronic and communication systems.

R&D Status and Trends in Nanoparticles, Nanostructured Materials, and Nanodevices in the U.S., published in December 1998, provides a broad perspective on U.S. R&D activities in nanoparticle synthesis and assembly; bulk materials behavior; dispersions and coatings; high-surface-area materials; functional nanostructures (devices); and relevant biological, carbon, and theory issues. The final report will cover nanotechnology in Japan, Western Europe, and Russia in addition to the United States.

The success of the nanotechnology workshop has established an ongoing project by NSF and WTEC called Nanobase. This unique program is a database of research centers, funding agencies, major reports and books, and conferences related to nanoscale S&T.[3]

In an effort to assess the extent to which WTEC, JTEC, and JTECH reports contribute to broad technical literature in their respective fields, we reviewed existing literature to discover where the reports are cited and how they are used. The team searched the Applied Science and Technology Index, FBIS, DIALOG, Compendex, and INSPEC in its efforts to find citations and references. However, only DIALOG, Compendex, and INSPEC produced results. These sources turned up 114 references to WTEC or JTEC in 56 industrial publications over 12 years. King Publications weeklies—*New Technology Week* and the NTIS *Foreign Technology Newsletter*—were the most likely publications to cite WTEC or JTEC activities with 31 and 9 citations, respectively. Twelve technical articles cited JTEC/WTEC/JTECH as a technical reference; four of these citations were found in the IEEE *Spectrum* and one was found in the *Czechoslovak Journal of Physics*. According to Phyllis Genther Yoshida, Director of the Asia-Pacific Technology Program at the DoC, one reason for this relatively low citation number is that

---

[3]Which can be found at **http://itri.loyola.edu/nanobase** (last accessed July 13, 1999).

much high-valued ISTI is only good for a short time and therefore is not cited often in bibliographies. But, as shown in Figure A.1, the frequency of WTEC and JTEC citations is generally on an upward trend, with a peak in 1996 with the optoelectronics report.

NTIS and ITRI distribute the WTEC and JTEC reports. Since its inception in 1989, ITRI has distributed more than 30,000 copies of its reports. The most widely distributed reports have been *JTEC/WTEC Annual Report and Program Summary* 1993/94 (3,468); *Display Technologies in Japan,* 1992 (2,800); *Satellite Communications System & Technology.* Vols. I and II, 1993 (2,000); *Knowledge-Based Systems in Japan,* 1993 (1,000); and *Electronics Manufacturing and Packaging in Japan,* 1995 (1,000).[4]

NTIS reported selling 83 WTEC reports from their library of seven different titles between 1995 and 1998. During the same period, NTIS distributed 140 JTEC reports out of a stock of 19 titles.

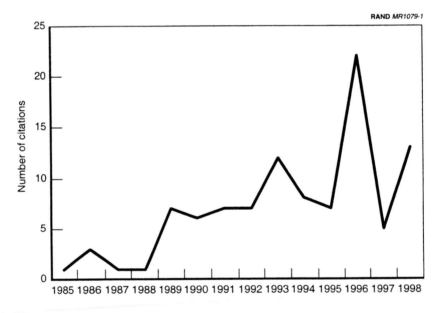

NOTE: Figures for 1998 reflect January–July only.

Figure A.1—Frequency of Citations

---

[4]Data provided by WTEC.

WTEC QUESTION RESPONSES

After we gathered information about citations of WTEC/JTEC literature, sponsorships, and subject matter, we moved on to contact sponsors, users, and likely users to ask them about the usefulness of the WTEC/JTEC information. Respondents fell into five groups: (1) government research directors and policymakers who sponsored WTEC research, (2) government scientists and engineers, (3) government analysts, (4) private-sector workshop participants, and (5) private-sector potential users. These five groups were asked the following questions about WTEC/JTEC:

1. Are you familiar with WTEC/JTEC? [If no, skip to question 8; if yes, continue.]

2. Have you used any of their products? Which of WTEC's/JTEC's different products or activities have you taken advantage of?

3. Are there particular studies that you recall as having been particularly good or influential? [If no, skip to question 6; if yes, continue.]

4. Why was this report(s) influential?

5. How did you use this report(s)?

6. Can you give a specific example of the way a report was used to change policy or affect research funding?

7. Overall, how would you judge WTEC's/JTEC's products?

Note that this set of questions did not constitute a formal survey; the list of participants is not a scientific sampling. Therefore, the results are indicative but not statistically verifiable.

Familiarity with WTEC/JTEC (Questions 1 and 2)

Of the 91 participants in the information-gathering phase, 67 were familiar with WTEC/JTEC studies. Obviously, sponsors and workshop participants had heard of WTEC/JTEC; when these participants are removed, about half of the respondents had heard of the WTEC/JTEC service or reports. Government sources were more likely to have heard of or used WTEC/JTEC at some time in the past. Private-sector sources that had not participated in the workshops were most likely to have never heard of WTEC/JTEC.

Of the 67 participants who had heard of or used WTEC/JTEC materials, most named a specific workshop or study that had been useful to them. Seven participants had heard of WTEC/JTEC, but could not recall a specific product or ever using a WTEC/JTEC product. Of the participants who named an influential study, most named a recent report (one issued within the last 5 years). The most frequently named subject of a report or workshop was optoelectronics, followed by satellites and nanotechnology.

The Usefulness of the Reports (Questions 3 and 4)

Participants were asked to report how influential the WTEC studies were in allocating research funding or making other decisions and why each was influential. We divided responses to the questions about how a WTEC/JTEC report affected individual or corporate decisionmaking into four broad categories:

- **Influential**—the report or workshop affected the direction of policy or R&D in government or industry.
- **Validating**—the report or workshop validated or confirmed what the user already knew about the topic.
- **Background**—the material served as background or information that fed into a larger stream of information being used to support the R&D process.
- **Not relevant**—the user did not find the information useful, either because the user was already aware of the information or because it was not directly relevant to his or her work.

Of the 67 participants in the information-gathering phase who were familiar with WTEC/JTEC, the largest number—25 users—cited the material as "influential." This group included workshop participants, who were, in general, the most positive about the usefulness of the WTEC/JTEC services. Eleven respondents reported that participating in the WTEC/JTEC workshop or reading a report validated what they had already suspected about the subject. Twenty-one respondents found the material useful as background information. Only five did not find the workshops or reports directly relevant.

Quality and Impact (Questions 5, 6, and 7)

Of the 54 participants who were familiar with the WTEC/JTEC products and the 11 sponsors of reports, 47 respondents found the services "excellent" (21) or "good" (26).[5] Only seven participants reported finding the material only "fair." These people reported they did not find the material of direct relevance to their work. Eleven participants said they could not comment on the overall quality of the materials because they did not have enough experience using them.

---

[5]Evaluation forms provided to workshop participants at the end of the WTEC/JTEC workshops generally rated the program as useful and good.

The impact of the WTEC/JTEC materials among those who reported that the materials influenced research direction or policy fell into four categories. Most frequently, participants said that the material led to a refined or redesigned R&D program. Close behind this were responses that indicated the WTEC/JTEC materials had led the users to modify their research emphasis. In several cases, WTEC/JTEC workshops and/or reports led government agencies or corporations to seek cooperative R&D activities with Japan in a specific area. In one case, a participant reported that the results of a WTEC/JTEC inquiry had influenced the creation of a U.S. R&D consortium.

Response by Type of User

When we break the responses down among the five types of user, we found the following results.

Government Research Directors and Policymakers. Among government research directors and policymakers who had sponsored WTEC activities, the majority of respondents reported that the WTEC activities were "influential" in their decisionmaking processes. The benefit of WTEC reports most frequently cited by the sponsors was timeliness: "It was timely, and this was key; the broader the interagency support, the more likely the reports were to have a bigger impact."

As might be expected, all the government research directors use a number of sources in addition to WTEC reports to keep abreast of developments in their field. No one source can completely meet their needs for information, but, when taken as a whole, all the sources together fairly well meet their needs for information. Support was high for government efforts to collect ISTI. When asked to comment overall, several points were made:

> The program should become more proactive and begin to monitor emerging competitors.
>
> We need to find out how well we did—are earlier [WTEC] reports still valid?
>
> Smaller targeted projects would be useful; short-term, getting the information out quickly is what is more useful.

Other government policymakers contacted for this study were generally supportive of WTEC but did not use the material in their work, so they could not comment directly on specifics. In general, these people felt they needed less technical detail and more "context" about the R&D situation in specific technologies or countries.

Government Scientists and Engineers. Government scientists and engineers contacted for this study were more likely to cite the WTEC materials as "validating" what they already knew or serving as background to their work.

This group relies heavily for technical information on the SRI S&T service. Only half of the 30 government scientists we contacted had heard of WTEC, while more than half use private services; of these, SRI is named most often.

Government Analysts. About half of the government analysts we contacted had heard of the WTEC products. Government analysts were the least likely to see the WTEC information as "influential." Government policymakers contacted for this study were generally supportive of WTEC but did not use the material in their work, so they could not comment directly on specifics. In general, these people felt they needed less technical detail and more "context" about the R&D situation in specific technologies or countries.

Private-Sector Workshop Participants. The private-sector workshop partici-pants were the most likely to see the WTEC activities as "influential." The materials and information were used in different ways, but the majority of these people viewed WTEC products as good or excellent. As one respondent noted of WTEC programs: "The programs provide enough information, and for a small company, it is good that government provides these workshops; other-wise, it would be too expensive."

Private-Sector Potential Users. About one-third of the private-sector potential users we contacted had heard of WTEC. The people who had heard of WTEC viewed the government material as good background.

WTEC-SPECIFIC RECOMMENDATIONS

Answers provided by our survey respondents have led us to recommend several program changes to be enacted in the WTEC program. These recommendation are outlined below[6]:

1. Expand the scope of mission beyond a national or regional focus to a technology-centered focus, identifying the state of the art of strategic tech-nologies worldwide.

2. Build an outreach plan to seek new WTEC customers and sponsors, specifi-cally among small businesses, and industry and government analysts.

3. Expand the scope of the individual technology assessments to include eco-nomic and trade information about the technology being examined.

4. Encourage WTEC managers to explore temporary or permanent alliances with other assessment services, such as SRI or MCC, to seek improved effi-ciencies in assessment or information dissemination.

------

[6]This list assumes that the government decides to continue the WTEC service in a capacity similar to that which exists now.

5. Increase face-to-face or real-time interaction among WTEC staff, panelists, and users by emphasizing seminars and workshops and creating on-line forums; decrease emphasis on published reports.

6. Increase the speed of information dissemination by employing information management tools and electronic media to both access and share vastly more information with a broader audience.

7. Build performance measures to support the expanded WTEC mission.

LIST OF WTEC REPORTS AND TABLES OF CITATIONS

Table B.1

WTEC Reports

Program	Report
JTECH	Panel Report on Computer Science in Japan (12/84) PB85-216760
JTECH	Panel Report on Opto- and Microelectronics (5/85) PB85-242402
JTECH	Panel Report on Mechatronics in Japan (6/85) PB85-249019
JTECH	Panel Report on Biotechnology in Japan (5/86) PB85-249241
JTECH	Panel Report on Telecommunications Technology in Japan (5/86) PB86-202330/XAB
JTECH	Panel Report on Advanced Materials (5/86) PB86-229929/XAB
JTECH	Panel Report on Advanced Computing in Japan (12/87) PB88-153572/XAB
JTECH	Panel Report on CIM and CAD for the Semiconductor Industry in Japan 12/88) PB89-138259/XAB
JTECH	Panel Report on the Japanese Exploratory Research for Advanced Technology (ERATO) Program (12/88) PB89-133946/XAB
JTECH	Panel Report on Advanced Sensors in Japan (1/89) PB89-158760/XAB
JTEC	Panel Report on High Temperature Superconductivity in Japan (11/89) PB90-123126
JTEC	Panel Report on Space Propulsion in Japan (8/90) PB90-215732
JTEC	Panel Report on Nuclear Power in Japan (10/90) PB90-215724
JTEC	Panel Report on Advanced Computing in Japan (10/90) PB90-215765
JTEC	Panel Report on Space Robotics in Japan (1/91) PB91-100040
JTEC	Panel Report on High Definition Systems in Japan (2/91) PB91-100032
JTEC	Panel Report on Advanced Composites in Japan (3/91) PB90-215740
JTEC	Panel Report on Construction Technologies in Japan (6/91) PB91 100057
JTEC	Program Summary (9/91) PB92-119429
JTEC	Panel Report on X-Ray Lithography in Japan (10/91) PB92-100205
WTEC	Panel Report on European Nuclear Instrumentation and Controls (12/91) PB92-100197
JTEC	Panel Report on Machine Translation in Japan (1/92) PB92-100239
JTEC	Panel Report on Database Use and Technology in Japan (4/92) PB92-100221
JTEC	Panel Report on Bioprocess Engineering in Japan (5/92) PB92-100213
JTEC	Panel Report on Display Technologies in Japan (6/92) PB92-100247
JTEC	Panel Report on Material Handling Technologies in Japan (2/93)PB93-128197

Table B.1—Continued

Program	Report
JTEC	Panel Report on Separation Technology in Japan (3/93) PB93-159564
JTEC	Panel Report on Knowledge-Based Systems in Japan (5/93) PB93-170124
NASA/ NSF	Panel Report on Satellite Communications Systems & Technology (7/93): Executive Summary PB93-231116 Vol. I. Analytical Chapters PB93-209815 Vol. II. Site Reports PB94-100187
WTEC	Monograph on Instrumentation, Control & Safety Systems of Canadian Nuclear Facilities (7/93) PB93-218295
JTEC/ WTEC	Annual Report and Program Summary 1993/94(3/94) PB94-155702
JTEC	Panel Report on Advanced Manufacturing Technology for Polymer Composite Structures in Japan (4/94) PB94-161403
ITRI	Benchmark Technologies Abroad: Findings from 40 Assessments, 1984-94 (4/94) PB94-136637
WTEC	Panel Report on Research Submersibles and Undersea Technologies (6/94) PB94-184843
JTEC	Panel Report on Microelectromechanical Systems in Japan (9/94) PB95-100244
WTEC	Panel Report on Display technologies in Russia, Ukraine, and Belarus (12/94) PB95-144390
JTEC	Panel Report on Electronic Manufacturing and Packaging in Japan (2/95) PB95-188116
JTEC	Monograph on Biodegradable Polymers and Plastics in Japan (3/95) PB95-199071
JTEC	Panel Report on Optoelectronics in Japan and the United States (2/96) PB96-152202
JTEC	Panel Report on Human-Computer Interaction Technologies in Japan (3/96) PB96-157490
WTEC	Panel Report on Submersibles and Marine Technologies in Russia's Far East and Siberia (8/96) PB96-199526
JTEC	Panel Report on Japan's ERATO and PRESTO Basic Research Programs (9/96) PB96-199591
JTEC/ WTEC	Panel Report on Rapid Prototyping in Europe and Japan: Vol. I, Analytical Chapters (3/97) PB97-162564 Vol. II, Site Reports (9/96) PB96-199583
WTEC	Panel Report on Advanced Casting Technologies in Japan and Europe 3/97) PB97-156160
WTEC	Panel Report on Electronics Manufacturing in the Pacific Rim (5/97) PB97-167076
WTEC	Panel Report on Power Applications of Superconductivity in Japan and Germany (10/97) PB98-103161
WTEC	Workshop Report on R&D Status and Trends in Nanoparticles, Nanostructured Materials, and Nanodevices in the United States (1/98) PB98-117914
WTEC	Panel Report on Electronic Applications of Superconductivity in Japan (7/98) PB98-150139
WTEC	Monograph on Use of Composite Materials in Civil Infrastructure in Japan (8/98) PB98-158215

Table B.2

JTEC Reports Cited in Industry Literature

Publication	Issue	Article	Subject
Computerworld	04/29/91 p. 101	"Learn More About Japanese Technology"	Adv. computing in Japan
NTIS Foreign Technology Newsletter	01/01/91 Vol. 91, No. 1	"New Report: JTEC Panel Report on Advanced Computing in Japan"	Adv. computing in Japan
Advanced Materials & Processes	08/93 Vol. 144, pp. 26–29	Vistasp M. Karbhari, and Diane S. Kukich "Polymer Composites Technology in Japan"	Adv. manuf. tech. for polymer composite structures in Japan
Composites & Adhesives Newsletter	06/93 Vol. 9, No. 3	"JTEC Announces Results of Japan Composites Processing Study"	Adv. manuf. tech. for polymer composite structures in Japan
New Technology Week	02/22/93 Vol. 7, No. 8	"Japan Reaches Parity in the Manufacture of Advanced Polymers"	Adv. manuf. tech. for polymer composite structures in Japan
Sensors & Instrumentation News	04/89	Marketing Sensors Conference Special Issue: "Political/Cultural Factors Impacting Sensor Companies," No. 4	Adv. sensors in Japan
Sensors & Instrumentation News	04/89	Marketing Sensors Conference Special Issue: "Technological and R&D Advances and Trends"	Adv. sensors in Japan
Sensor Trends International	01/89 Vol. 3 , No. 1	"Japan Widens Research Lead"	Adv. sensors in Japan
Sensors & Instrumentation News	11/88 Vol. II, No. 10	"New Products-JTECH Lauds Japanese Sensor Research"	Adv. sensors in Japan
International Journal of Materials & Product Technology	01/95 Vol. 10, No. 1–2	Vistasp Karbhari "New Product and Process Development Methods as Applied to Polymer Matrix Composites: A Perspective Influenced by a JTEC Study"	Adv. manuf. tech. for polymer composite structures
Equipment & Materials Update	04/94 Vol. 94, No. 4	"Japanese Advanced Manufacturing Technologies for Polymer Composite Structures"	Adv. manuf. tech. for polymer composite structures

Table B.2—Continued

Publication	Issue	Article	Subject
Chem Eng. News	11/17/86 Vol. 17, pp. 8–10	Richard J. Seltzer, "Japan Aims for World Leadership in Advanced Polymers by 1990s"	Advanced materials
Chemical Week	04/23/86 Vol. 64, No. 46, pp. 35–38	Joseph F. Dunphy, Shota Ushio "Japan's Advances in Polymers Pose a Challenge"	Advanced materials
New Technology Week	07/20/92 Vol. 6, No. 29	"The Japanese Technology Evaluation Center Has Released a New Report Entitled Bioprocess Engineering In Japan"	Bioprocess engineering in Japan
Manufacturing News	04/15/96 Vol. 3, No. 8	"Japanese Foundries Face Many of the Same Problems"	Casting
Computer-Aided Design	10/15/89 Vol. 10, No. 10	"Current Practice in Japan Advanced Manufacturing Technology"	CIM & CAD for semiconductor ind.
Computer Integrated Manufacturing	08/15/89 Vol. 10, No. 8	"Current Practice in Japan Advanced Manufacturing Technology"	CIM & CAD for semiconductor ind.
Composites & Adhesives Newsletter	05/92 Vol. 8, No. 4	"DOE Lab Differences Again Surface"	Composites
Automation in Construction	05/92 No. 1, pp. 27–34	R. L. Tucker "Japanese Construction Industry"	Construction technologies
NTIS Foreign Technology Newsletter	11/13/90 Vol. 90, No. 46	"Panel Evaluates Japanese Construction Technologies"	Construction technologies
NTIS Foreign Technology Newsletter	11/26/91 Vol. 91 , No. 48	"New Report JTEC Panel Report on Construction Technologies in Japan"	Construction technologies in Japan
Telecommunications Policy	04/92 Vol. 16, No. 3, pp. 259–276	Martin Fransman "Japanese Failure in a High-Tech Industry"	Database use and technology
High Performance Computing	05/26/98 Vol. 7, No. 21	Mark Crawford "Japan Prepares to Play in Future Digital World"	Digital information organization in Japan
New Technology Week	05/18/98 Vol. 12, No. 20	"Japan Prepares to Play in Future Digital World"	Digital information organization in Japan
Defense & Aerospace Electronics	09/7/92 Vol. 2, p. 6	"Study Examines Display Panel Competition; Japanese Technology Evaluation Center's Report, 'Display Technologies In Japan'"	Display technologies in Japan

Table B.2—Continued

Publication	Issue	Article	Subject
Computerworld	02/10/92 pp. 31	Michael Alexander, CW Staff "Advanced Technology: U.S. to Topple Japan in Am-LCD Market?"	Display technologies in Japan
New Technology Week	07/29/96 Vol. 10, No. 31	"Advice from Burgeoning Singapore"	Electronic manuf. in the Pacific Rim
New Technology Week	07/29/96 Vol. 10, No. 31	"Maturing Asian Electronics Industry Threatens to Capture More of World's Sales, Investment"	Electronic manuf. in the Pacific Rim
Electronic Buyers' News	04/17/95 p. 48	Jack Robertson "Manufacturing & Test: New PCB Technology Sought"	Electronic manuf. and packaging in Japan
SMT Trends	01/95 Vol. 11, No. 12	"From the Editor: New Year Brings New Challenges for Electronics Manufacturing"	Electronic and manuf. and packaging in Japan
New Technology Week	01/18/94 Vol. 8, No. 3	"Packaging, Miniaturization Drive Japan's Electronics Market Strategy"	Electronic manuf. and packaging in Japan
Proceedings of the 16th IEEE/CPMT International Electronics Manufacturing Technology (IEMT) Symposium	09/12–14/94 Part 1, Vol. 1	Michael Pecht, Michael Kelly, William Boulton, John Kukowski, Gene Meieran, John Peeples, Rao Tummala "JTEC Panel Report on Electronic Manufacturing and Packaging in Japan"	Electronic manuf. and packaging in Japan
SMT Trends	02/1/96	"Notes on Japan's Electronics Industry"	Electronic manufacturing in Japan
NTIS Foreign Technology Newsletter	03/06/90	"ERATO Program Looks at Superbugs, Atomcraft and Bimolecular Machines"	ERATO
New Technology Week	12/18/95 Vol. 9, No. 50	Mark Crawford "ERATO Sparks Japan's Basic Research Effort"	ERATO and PRESTO
Research-Technology Management	11 & 12/96 Vol. 39, No. 6, pp. 3–4	Phyllis Genther Yoshida "Japan Pursues Creative Science/Technology"	ERATO and PRESTO basic research programs
New Technology Week	11/25/91 Vol. 5, No. 47	"Is the Race Ending to Commercialize Flat Panel Displays?"	Flat panel display tech.

Table B.2—Continued

Publication	Issue	Article	Subject
Electronic News	04/12/93 Vol. 39, No. 1958, p. 14	David Kellar "Flat Panel Technology Race Lures More Entries" (Flat Panel Displays) (1991)	Flat panel displays
Wall Street Journal	12/13/91 p. B5A(W), p. A9B(E)	Eduardo Lachica "Researchers of Flat-Panel Screens See Opportunity to Match the Japanese" (Active-Matrix Liquid Crystal Displays) Fri. ed.	Flat panel displays
High Performance Computing	06/8/98 Vol. ,7, No. 23	"Japan, Asia Threaten U.S. Lead in Hard Disk Drives"	Future of data storage tech.
High Performance Computing	06/1/98 Vol. 7, No. 22	"Asian Manufacturers Dominate Optical Storage"	Future of data storage tech.
New Technology Week	06/1/98 Vol. 12, No. 22	"Japan, Asia Threaten U.S. Lead in Hard Disk Drives"	Future of data storage tech.
New Technology Week	06/10/96 Vol. 10, No. 24	"HCI Technology Growing in Importance to Market"	HCI
SMT Trends	05/27/96 Vol. 13, No. 5	"PCB DESIGN & CAD: EMPF Working on Design for Manufacturing"	HCI
Manufacturing News	05/15/96 Vol. 3, No. 10	"Leads in Computer Interfaces"	HCI
NTIS Foreign Technology Newsletter	03/20/90 Vol. 90, No. 12	"JTEC Panel Report on High Temperature Superconductivity in Japan"	High temp. supercdvty. in Japan
Science	09/22/89 Vol. 245, No. 4924, p. 1331	Robert Pool "Japan: Superconductor Hopes Drop"	High temp. supercdvty. in Japan
Federal News Service	12/6/90	"Hearing of the Joint Economic Committee Subject: Japan's Economic Challenge: Advanced Technology Issues Chaired By: Representative Lee H. Hamilton (D-In) Witnesses: William F. Finan Principal Technecon, Inc. Kenneth Flamm Senior Fellow Brookings Institute Martha Harris Director Office Of Japan Affairs National Research Council Room 2261, Rayburn House Office Building"	JTEC
New Technology Week	08/4/97 Vol. 11, No. 31	"Sleeping Tiger? Think Again! WTEC Calls Inattention to Japan's Might Naive"	JTEC commentary
Communications of the ACM	01/94 Vol. 37, No. 1	"Knowledge-Based Systems in Japan; Japanese Technology Evaluation Center Panel Report on Japan's Expert Systems"	Knowledge-based systems in Japan

Table B.2—Continued

Publication	Issue	Article	Subject
New Technology Week	05/03/93 Vol. 7, No. 18	"Need Sparks Japanese to Push State of Art in Material Handling"	Material handling
Material Handling Engineering	07/92 Vol. 47, No. 7, p. 4	Gene F. Schwind "Japan Jumps for MH Technology; Materials Handling"	Material handling in Japan
American Metal Market	06/10/85 Vol. 93, p. 11	Gerry Khermouch "Japan Outstrips US in 'Smart-Machine' Technology: Commerce Dept.'s draft; Island Nation Trailing Only in Software AI but Is Dedicated to Making Drive in That Sector"	Mechatronics
Technology Alert	03/1/98 Vol. 98, No. 2	"Microelectromechanical Systems Developments in Japan"	MEMS
Electronic Engineering Times	11/29/93 No. 774, p. 1	Brian Santo "Japan Gears up for Micromachines"	MEMS
New Technology Week	11/22/93 Vol. 7, No. 47	"Leads MEMS Field, But Japan's on the Move"	MEMS
Technology Review	05/98 Vol. 101, No. 3, p. 30.	David Rotman "Big Money for a Small World"	Nanotech.
Fine Particle Technology News	05/98 Vol. 1, No. 4.	"Panel Reviews Global R&D Status"	Nanotech.
High Performance Computing	02/23/98 Vol. 7, No. 8	"Lucrative Nanotech Derby Still Wide Open"	Nanotech.
Science	02/20/98 Vol. 279, No. 5354, p. 1121a	"Nanotech Gold"	Nanotech.
New Technology Week	02/17/98 Vol. 12, No. 7	"Lucrative Nanotech Derby Still Wide Open"	Nanotech.
High Performance Computing	01/5/98 Vol. 7, No. 1	"NSF Leads Nanotech Coordination Effort"	Nanotech.
Federal Technology Report	05/22/97 p. 4	Neil Macdonald "Agencies Review Diverse Blend of Nanotechnology R&D Programs"	Nanotech.
Nuclear News	04/91 p. 19	"Recently Published: JTEC Panel Report on Nuclear Power in Japan"	Nuclear power in Japan
NTIS Foreign Technology Newsletter	12/18/90 Vol. 90, No. 51	"New Report: JTEC Panel Report on Nuclear Power in Japan"	Nuclear power in Japan
EMedia Professional	01/98 Vol. 11, No. 1, p. 60	Dana J. Parker "The Many Faces of High-Density Rewritable Optical"	Optoelectronics in Japan and in the US

Table B.2—Continued

Publication	Issue	Article	Subject
Fiber Optics News	06/17/96 Vol. 16, No. 25	"JTEC Report: Japanese Optoelec- tronics Industry Tops $40 Billion; U.S. $6 Billion"	Optoelectron- ics in Japan and the U.S.
M2 Presswire	05/8/96	IGI Consulting "Optoelectronics in Japan and the United States"	Optoelectron- ics in Japan and the U.S.
Manufacturing News	04/1/96 Vol. 3, No. 7	"In Optoelectronics, Japan Holds Most but Not All of the Cards"	Optoelectron- ics in Japan and the U.S.
New Technology Week	04/14/97 Vol. 15, No. 11	"U.S. Facing Japanese Challenge in Superconducting Electronics"	Power apps. of supercon- ductivity in Japan and GR
Defense Week	08/12/96 Vol. 17, No. 33	"Japan Lead in Superconductors Vulnerable to Germany"	Power appsof supercon- ductivity in Japan and GR
New Technology Week	08/5/96 Vol. 10, No. 32	"U.S. Lead in Superconductors Vul- nerable to Germany, Japan"	Power appsof supercon- ductivity in Japan and GR
Composites Industry Monthly	01/93	"Japanese Technology Evaluation Center Overview at U.S. Workshop"	Program sum- mary
New Technology Week	07/28/97 Vol. 11, No. 30	"Rapid Prototyping: Europe, Japan Challenge U.S. Lead"	Rapid proto- typing
Manufacturing News	04/15/96 Vol. 3, No. 8	"Rapid Prototyping in Japan, Europe and the United States"	Rapid proto- typing
Materials Technology	Mar-Apr 1996 Vol. 11, No. 2, pp. 62–66	Michael J. Dehaemer "Highlights of the Workshop on Rapid Prototyping in Japan and Europe" Source: Matrice Tech Ltd. Sidcup Engl.	Rapid proto- typing
New Technology Week	08/02/93 Vol. 7, No. 31	"Submersibles in Ex-USSR Eye Openers for Westerners"	Rsrch sub- mersibles & undersea tech.
Sea Technology	12/94 Vol. 35, p. 23	Richard F. Burns "Research Submersibles, undersea technologies (in W W. Eur. and former Sov. Union"	Rsrch sub- mersibles & undersea tech.
Defense Week	12/22/97 Vol. 18, No. 50	"Satellite Communications: U.S. Dominance at Risk?"	Satellite comm. systems & tech.

Table B.2—Continued

Publication	Issue	Article	Subject
Defense Week	12/22/97 Vol. 18, No. 50	"Satellite Communications: U.S. Dominance at Risk?"	Satellite comm.systems & tech.
High Performance Computing	12/15/97 Vol. 6, No. 12	"Satellite Communications: U.S. Dominance at Risk?"	Satellite comm.systems & tech.
New Technology Week	12/8/97 Vol. 11, No. 48	"Satellite Communications: U.S. Dominance at Risk?"	Satellite comm.systems & tech.
Satellite Week	02/15/93 Vol. 15, No. 7	"Lead in Mobile Satellite Services Will Disappear, Study Says"	Satellites comm. systems & tech.
New Technology Week	02/08/93 Vol. 7, No. 40	"Satellites: Another U.S. Industry Faces Decline"	Satellites comm. systems & tech.
New Technology Week	12/23/91 Vol. 5, No. 51	"In Advanced Chipmaking, United States and Japan Are Running Neck to Neck"	Semiconductors
New Technology Week	05/10/93 Vol. 7, No. 19	"Japan Drawing Bead on U.S. in Membranes?"	Separation technology
New Technology Week	07/29/96 Vol. 10, No. 31	"Flat-Panel Surge by Korea, Taiwan Raising More Questions for U.S."	South Korea electronics industry
Solid State Technology	09/96 Vol. 39, No. 9, p. 76	Martin Peckerar South Korea: A New Tiger Faces New Technology Issues	South Korean electronics industry; JTEC
NTIS Foreign Technology Newsletter	10/23/90 Vol. 90, No. 43	"JTEC Panel Report on Space and Transatmospheric Propulsion Technology"	Space propulsion in Japan
New Technology Week	08/13/90 Vol. 4, No. 33	"Japanese Take Minimalist Approach to Ambitious Space Robotics Program"	Space robotics in Japan
Sea Technology	12/96 Vol. 37, p. 23	"Submersibles, Marine Technologies"	Submersibles & marine tech.
Marine Technology Society Journal	Spring 1996 Vol. 30, pp. 71–72.	Brad Mooney, Richard Seymour "WTEC Panel Survey Russian Maritime Technologies"	Submersibles & marine tech.
BNA International Business & Finance Daily	08/04/97	"Despite Recession, Japan Beats Western Nations In R&D Spending, Conference Told"	Workshop
NTIS Foreign Technology Newsletter	06/04/91 Vol. 91, No. 23	"Panel Evaluates X-Ray Lithography in Japan"	X-ray lithography

Table B.2—Continued

Publication	Issue	Article	Subject
Chilton's Electronic News	02/25/91 Vol. 37, No. 1849, p. 27	Jack Robertson "Japan: X-ray may Wait for 1g Dram; X-Ray Lithography, Dynamic Random Access Memory; Gigabyte"	X-ray lithography
Private Equity Week	07/14/97 Vol. 4, No. 28, pp. 1ff	"High-Tech Developer Looks to the Next Wave"	
Science	05/3/96 Vol. 272, No. 5262, p. 645	Dennis Normille "New Faculty Grants Program Expands Role Of STA Agency; Science and Technology Agency"	
New Technology Week	05/30/95 Vol. 9, No. 22	"Study Faults U.S. Daring in High-Volume Products"	
Warfield's Business Record	07/7/93 Vol. 8, No. 30, Sec. 1, p. 13	Tom Johnson "Tracking Japan's Growing Strength in Development of High Technology"	
New Technology Week	04/26/93 Vol. 7, No. 17	"Loyola Furthers Global Tech Agenda"	
Inside DOT & Transportation Week	04/23/93 Vol. 4, No. 16	"Japanese Technology Gains Studied at Loyola College"	
New Technology Week	09/21/92 Vol. 6, No. 37	"U.S. Firms Slow to Tap Japanese Technical Data"	
New Technology Week	06/29/92 Vol. 6, No. 26	"Assesses Japan's Fifth Generation Computer Project"	
New Technology Week	01/06/92 Vol. 6, No. 1	"The Japan Technology Evaluation Center"	
New Technology Week	05/28/91 Vol. 5, No. 22	"JTEC Takes Measure of Japanese Biotechnology"	
Database	04/91 Vol. 14, No. 2, pp. 105–107.	R. R. Dueltgen "Access to Japanese Technical Information"	
NTIS Foreign Technology Newsletter	09/18/90 Vol. 90, No. 38	"Panel Evaluates Japanese Materials R&D"	
Superconductor Week	04/02/90	"Japanese Dominate Digital Josephson Technology"	
Superconductor Week	03/19/90 Vol. 4, No. 12	"DOE Laboratory, Headquarters Differences Again Surface"	
Automation News	01/88 Vol. 2, No. 1	"US Reports on High Technology R&D in Japan Manufacturing"	
IEEE Spectrum	10/85 Vol. 22, No. 10, pp. 46–52	Trudy E. Bell " Japan Reaches Beyond Silicon"	

Table B.3

JTEC Reports Cited as References

Publication	Date	Author(s) and Article Title	Report Cited
Quality and Reliability Engineering International	Jan.–Feb. 1997 Vol. 13 , No. 1, pp. 3–8	N. Kelkar, A. Dasgupta, M. Pecht, I. Knowles, M. Hawley, D. Jennings "'Smart' Electronic Systems for Condition-Based Health Management" (reprint)	MEMS 1994
Science	Nov. 1, 1996 Vol. 274, No. 5288, pp. 736–737	David Larbalestier "Superconductor Flux Pinning and Grain Boundary Control"	Superconductivity in Japan and GR
Manufacturing News	Aug. 5, 1996 Vol. 3, No. 15	"Words, Recent Manufacturing Reports"	Advertisement for Manuf. Reports
Computerworld	July 3, 1995 p. 66	Michael Fitzgerald "Summer Breezes"	Manufacturing Reports to Read
Issues in Science and Technology	Fall 1994 Vol. 11, No. 1, pp. 27–32	Kenneth S. Flamm "Flat-Panel Displays: Catalyzing a U.S. Industry"	Display Technology (Recom. Reading)
Proceedings of the IEEE	Apr. 1994 Vol. 82, No. 4, pp. 499–509	Lee Tannas "Evolution of Flat-Panel Displays"	Display Technology 1992
IEEE Expert-Intelligent Systems & Their Applications	Feb. 1994 Vol. 9, No. 1	J. Hendler "Beyond the 5th Generation Parallel AI Research in Japan"	JTEC 1992 Study (not specified)
IEEE Communications Magazine	May 1993 Vol. 31, No. 5, pp. 88–94	D. Stokesberry, S. Wakid "ISDN in North America"	Telecom Report 1986
Journal of Vacuum Science & Technology	Nov.–Dec. 1992 Vol. 10, No. 6	D. Fleming , J. Maldonado, M. Neisser "Prospects for X-Ray Lithography"	X-ray Lithography 1991
Trends in Biotechnology	1989 Vol. 7, No. 9, pp. 227–229	D. L. Oxender, C. F. Fox "Barriers to United-States Industrial Biotechnology Research Consortia"	JTECH Report 1985 and 1988 (Biotechnology)
Czechoslovak Journal of Physics	1987 Vol. 37, No. 2, pp. 257–272	J. Vlachy "World Trends, Publication Output, Research Fronts and Highly Cited Papers in Optics, Lasers and Quantum-Electronics"	Opto and Micro Electronics 1985
IEEE Spectrum	1986 Vol. 23, No. 6, pp. 47–52	"Assessing Japan's Role in Telecommunications"	Telecom Report 1986

Table B.4

Cited Conferences

Conference	Date and Place	Subject	Speaker and Topic
100 Casting Congress and CastExpo	April 20–23, 1996 Philadelphia, Penn.	Casting Technologies	H.W. Hayden United States Japanese R&D on New Cast Alloys and Materials
Semiconductor Equipment and Materials International: Semicon West Conference	July 1994 San Francisco, Calif.	Electronic Manuf. and Packaging in Japan	R. Tummala Japanese Packaging Roadmap (Based on NSF-Sponsored JTEC Study): Packaging Industry: Roadmap to the Future (SEMI, pp. 124–141)
Inside Conference: Advanced Display Technologies in Russia, Ukraine and Belarus Workshop	February 1994 Arlington, Va.	Advanced Display Technologies in Russia, Ukraine and Belarus	M. DeHaemer Introduction to WTEC (pp. 1–13)

SERVICES PROVIDING SOURCES OF ISTI INFORMATION

Many of the services available in the late 1990s were established between 1978 and 1988 during the height of the Japanese challenge years. The ISTI services grew up in a relatively random fashion. For example, in a 1990 report, the GAO noted that 62 different offices and divisions within 6 agencies monitored foreign technology. In its review, the GAO found that the departments of Commerce, Defense, Energy, and State, as well as NASA and the NSF, accounted for a significant portion of foreign technology information. (The Central Intelligence Agency [CIA] did not cooperate with the GAO, so the GAO study review could not cover the FBIS and other CIA activities.) Nevertheless, the GAO "could find no central list of agencies that monitor foreign technology." And despite what looked like broad coverage, "the GAO also found that coordination among monitoring agencies is limited." (GAO, 1990, p. 2.) A subsequent GAO study failed to find an increase in coordination.[1] An entrepreneurial analyst or researcher had to seek information from a large number of disparate sources.

This appendix describes the most notable unclassified services in the public and private sectors, as well as journals and publications that cover this kind of material.

GOVERNMENT SOURCES OF ISTI

One way to categorize government sources of ISTI information is by the type of information the various public-sector services provide. We have distinguished between three categories:

- Technical literature, such as data, drawings, and research findings, translated, when necessary, from foreign news sources were produced for gov-

---

[1]In addition to the unclassified services, the government has developed, maintained, and sometimes abandoned classified services providing similar information over the years. These services are not described in detail in this report. However, GAO (1990) and GAO (1992) provide descriptions of both unclassified and classified ISTI services in operation at the end of the 1980s and early 1990s.

ernment analysts seeking to create reports and assessments that would feed acquisition, technical, and policy-level decisions.

- Snapshot assessments were aimed at supporting government scientists or laboratory directors who could use the information to "benchmark" their progress against that of other countries.

- Full-scale assessments produced by the OTA and like offices were written to answer for the policymaker facing specific research funding allocation or programmatic decisions.

Some examples of U.S. government services are described below.

Literature Reviews and Monitoring

FBIS. The FBIS, initially created by the Federal Communications Commission in 1941 to monitor Japanese, German, and Italian broadcasts, became a part of the CIA's Directorate of Science and Technology after the Second World War. The FBIS was tasked to provide translations of foreign government and academic journals and news sources on a broad range of topics; S&T information became an increasingly important focus for the FBIS as the Japanese challenge changed the demand for information.

The translated S&T abstracts or full text are provided to government analysts via the Joint Publications Research Service reports and the FBIS daily reports. The former focuses on social science and technical journals, while the FBIS daily reports emphasize radio broadcasts and other news publications. Information that cannot be shared beyond the government for copyright and access reasons is packaged as "For Official Use Only" and is not available outside the government. Much of the FBIS information is not publicly available.

In the past, some of the FBIS information that was not protected by copyright, or when the U.S. government made a special royalty arrangement, was made available to the public through unclassified publications. These publications are no longer available from FBIS. Currently, FBIS information that is made available to the public is provided through the NTIS service called "World News Connection" on the World Wide Web at **http://wnc.fedworld.gov** and in paper copies that can be ordered through NTIS.

JTLP. By the mid-1980s, it was clear that the Japanese were transferring and applying U.S. technical knowledge to create Japanese products. Yet American companies knew very little about Japanese technological capabilities. Access to, and use of, Japanese technical literature was nearly unheard of, partly because of the bottleneck posed by identifying and translating information from Japanese into English (Hill, 1987).

To address the situation, Congress passed the Japanese Technical Literature Act of 1986 (Public Law 99-382), creating the JTLP at DoC. Congress directed JTLP to monitor, acquire, and translate Japanese technical publications and make them available to the research and policy community. In responding to this mandate, JTLP initiated several programs to disseminate technical information. It published the Japanese Technical Literature Bulletin, a quarterly report on Japanese high-technology development; the Directory of Japanese Technical Resources in the United States; the Directory of Japanese Technical Reports; and other topical assessments. JTLP funding was approximately $400,000 per year for the first six years.

Snapshot Assessments

Two examples of services providing snapshot assessments are Science and Technology Reporting Information Dissemination and Enhancement (STRIDE) and Worldtec.

STRIDE. Executive Order 12591, "Facilitating access to S&T," issued by President Ronald Reagan in April 1987, directed a number of changes to the way international S&T information is collected and disseminated. Specifically, the executive order states that

> The Secretaries of State and Commerce and the Director of the National Science Foundation shall develop a central mechanism for the prompt and efficient dissemination of science and technology information developed abroad to users in Federal laboratories, academic institutions, and the private sector on a fee-for-service basis.[2]

In response to Executive Order 12591, the relevant departments, led by State, established STRIDE. To implement STRIDE, the Department of State established a code (actually, the code was just "STRIDE") to be placed along with other codes on the dissemination line for cables being issued by U.S. Embassies overseas.[3] STRIDE information must be unclassified to be included in the database. In the early days, the NTIS incorporated this information into a newsletter called "Foreign Technology—An Abstract Newsletter." In the mid-1990s, the name of the newsletter changed to the "Foreign Technology Update." Since 1997, STRIDE data and other international S&T information have been compiled in an NTIS database called "Worldtec."

---

[2]52 FR 13414, 3 CFR, 1987 Comp., p. 220, Sec. 4. "International Science and Technology," subsec. (c).

[3]The coding system provides a routing indicator; the routing codes show whether a cable would be of interest to economic officers, trade officers, political officers, and so on, within the U.S. government.

Worldtec. Worldtec is accessible to government, academic institutions, and the private sector through a subscription service on the NTIS World Wide Web server (**http://www.ntis.gov**). The Worldtec/STRIDE reports are "snapshots" of a scientific or technical development in another country. A report tends to be along the lines of a trip report based on the visit of an embassy officer to a science or technology center in a specific country. There is no requirement for embassy officers to report on any specific technology, and there is no regularity to the Worldtec/STRIDE reporting.[4]

Since the vast majority of cable traffic coming from U.S. Embassies overseas is classified, the amount of reporting that can be slotted for STRIDE has been very small. In a limited number of cases, relevant cable traffic is declassified for use in Worldtec. In addition to cable traffic, some additional data are included in Worldtec, such as the freely available short reports that are sent out over the Web by the Asian Technology Information Project. According to NTIS, Worldtec is not widely used; in fact, the NTIS database has fewer than 600 records and fewer than 150 subscribers.

Analytic Reports on Technology

The OTA, ONR, and other agencies provide examples of services providing analytic reports on technology.

OTA. Created in 1972 and disbanded in 1995, the Congress's OTA had an initial mandate to provide early indications of the probable beneficial and adverse impacts of the applications of technology. OTA's mandate grew over time to encompass the analysis of a broad range of S&T policy issues. The demand for technology assessment expanded as Congress faced increasingly complex scientific and technical issues. Joining this demand was public concern about the risks of adopting new technologies following a series of domestic alarms and disasters: pollution in residential and agricultural areas, controversy over the development of supersonic transport planes, and the thalidomide tragedy.

OTA focused almost exclusively on issues of concern at the policy level within the United States. Only infrequently did Congress request a study of OTA that included a global technology component. Of more than 750 OTA assessments released by the governing Technology Assessment Board, only 63 had, as a principal purpose, the assessment of some global issue. Even among these reports, only 12 appear to have made a point of assessing foreign technology capabilities. OTA was not a primary source of ISTI or global technology assessments.

---

[4]The Department of State also periodically issues some longer reports on important industries in specific countries. The subject of the report would depend on the country in question.

ONR. The ONR maintains ISTI collection offices in London and Tokyo to feed material continually to a team of researchers tracking foreign technical capabilities. Although tasked to provide information on those technologies with potential naval application, ONR reads this mandate broadly and produces assessments on a range of important technologies. While a number of these reports are classified, ONR has also produced high-quality, unclassified assessments of foreign technology. Examples of the types of report ONR Asia has issued and that are currently available on the World Wide Web at **http://www. onr.navy.mil/onrasia/** include: *Trends of Composites Research in Japan*, May 8, 1998; *Current Trends of Polymer Research in India*, May 1, 1998; *Ceramics Research in Asia*, December 9, 1997, and *Functional Polymers Research in China*, August 18–September 8, 1996.

ONR's European office maintains a comprehensive database of newsletters, reports, and updates on S&T, which is available at **http://www.ehis. navy.mil:80/onrnews.htm**. The news briefs include a wide range of topics, from high temperature superconductivity to optoelectronics. Furthermore, the ONR's European office is also developing a scientific and technical conference database allowing individuals to search by keyword on forthcoming conferences.

Other Agencies. Individual agency offices and units have also undertaken projects to meet particular needs. The various government agencies collect or undertake technology assessments to help them make decisions about their own R&D activities or foreign acquisition. The NSF collects information about scientific developments and policy abroad. Subjects include astronomy, chemistry, materials, mathematics, physics, and computing. The information is used to evaluate grant proposals. Other agencies, such as the Department of Defense and DoE, assess technologies for acquisition or use in research. Many of these reports are buried within the agencies themselves and do not circulate widely to industry or academe. Examples include the following:

- *Benchmark Tests on Digital Equipment Corporation Alpha AXP 21164-based Alpha server*, by H. J. Wasserman, a 1996 DoE study

- *Comparative Nuclear Engine System Simulation*, by D. G. Pelaccio and others, a 1997 NASA study

- *Meeting the Challenge: U.S. Industry Faces the 21st Century: The Chemical Industry*, by Allen Lenz and John Lafrance, a 1996 DoC study.

NONGOVERNMENT SOURCES OF ISTI

Collecting unclassified ISTI and assessing foreign capabilities started out as the purview of government. Certainly, information of this kind has a "public goods" component—it can be openly collected and shared among many people at the same time, and it is difficult to prevent others from using the product. These features make it less likely that a private supplier will emerge to provide the service. Open and shared information has features of what might be called a "commodity"—a commonly available good that aids commerce. Over time, private industry and some government agencies found that they needed more "strategic" commercial information—information that is provided solely to the "customer" for its use in decisionmaking within a limited period. Often, information that starts out being of strategic importance is then more widely shared and becomes part of the pool of general knowledge. It is the timeliness of the information that often gives it a strategic character.

Private services are also a phenomenon of the Japanese challenge period. They were launched mostly in the late 1970s and early 1980s to collect targeted and strategic information on foreign S&T. These services also produce assessments of the competitive capabilities and, at times, the market position of foreign competitors. The services described below were all created, at least in part, with support from government agencies. Other sources, supported wholly by private-sector consumers began to develop as well, including the following:

- Dataquest (McGraw-Hill), providing market intelligence to industry through an international network of research, information, custom consulting, and analysis. Dataquest offers market intelligence concerning more than 25 specific information-technology markets from an international perspective.

- Frost & Sullivan Market Intelligence Service, providing technical and market assessments about 10 sectors of the economy from agriculture to electronics to transportation. Frost & Sullivan also sponsors workshops on technical and business-related topics.

- Booz·Allen & Hamilton, providing targeted, timely information about clients' technology challenges and global market conditions.

In general, services supporting the private sector, because they are providing targeted and quite specific information, charge a hefty fee and are quite expensive compared to the government-provided services.

SRI

SRI International (**http://www.sri.org**) is a research, technology-development, and consulting organization that began technology-tracking services under contract to the NSF in 1978. Through its Washington-based Science and Technology Policy Program, the company provides research and analysis on international scientific and technical issues to the NSF and other government agencies, including the U.S. Agency for International Development, the DoE, NASA, the DoC, the ONR, and the State Department. The assessments include a variety of topics, ranging from analyses on the programs and capabilities of individual nations, such as Japan, Brazil, and South Korea, to examinations of the impact of foreign science and engineering students in the United States.

MCC

In response to the rise in Japan of targeted high-technology R&D consortia, MCC (**http://www.mcc.com**) was established in 1983 as America's first R&D consortium composed of computer, semiconductor, and electronic manufacturers. At its inception, it also had some support from government. Buttressed by the passage of the National Cooperative Research Act (PL 98-462) in 1984 allowing basic research cooperation among firms within the same industry, its operation heralded a new era in cooperation among U.S. high-technology Industries.[5] Through MCC's International Liaison Office, the consortium provides its members with a monthly technical report on foreign technology, tracks R&D in emerging electronic and information technologies, and provides site visits to foreign laboratories. The liaison office has established two databases, one that tracks Japanese technical journals, publications, and conferences and another that provides full-text research activities in Japan and Europe. MCC has also developed a database that includes technical reports from JTEC, ONR, the ATIP, and the MIT-Japan Program.

ATIP

The ATIP (**http://www.atip.org**), founded in 1990 by Wally Lopez and Dr. David Kahaner after he left the service of the ONR-Tokyo, provides information and analysis on technology developments in Asia. ATIP provides private consulting services; translates Asian technology documents; and, through its electronic alert system, informs members of upcoming reports, seminars, and conferences. In conjunction with DARPA, ATIP funds a Flat Panel Display program and a Micro Electro-Mechanical Systems (MEMS) program.

FOREIGN SOURCES

One of the major sources of scientific and technical information is the foreign government itself. The Japan Information Center of Science and Technology (JICST), the British Library, and the French periodical *Technologies "France"* from the *Agence pour la Diffusion de l'Information Technologique* (ADIT) are just a few examples of foreign government repositories.

---

[5]The act amended U.S. antitrust law, allowing companies to enter joint ventures for R&D purposes at the precompetitive level.

Journals and Publications

In addition to the various government and private-sector services, dozens of groups produce reports and articles with information about S&T developments, some of which contain information about activities in other countries. Examples of this kind of material are IEEE's *Spectrum*; the Industrial Research Institute's journal, *Research-Technology Management*; and the American Chemical Society's *Chemical Abstracts.*

Scholarly journals, usually published by universities, also contain important information for researchers tracking S&T around the world. Examples of these are MIT's *International Journal of Robotics Research,* Harvard's *Information Technology Quarterly*, and the University of Pennsylvania's *The Scientist.*

Technical magazines also provide reporting on technology developments. They provide some focus on Japanese and European activities, rarely covering other parts of the world. Magazines like these include *Technology Review, Scientific American*, and *Communications of the ACM.* Some news weeklies, such as *The Economist*, also report on S&T developments.

EXAMPLES OF SCIENTIFIC AND TECHNICAL ASSESSMENTS FROM SRI, MCC, ONR (ASIA AND EUROPE), AND ATIP

SRI

The following are recent reports from SRI, which is based in Washington:

- *Assessment of NSF's Role in Neuroscience Research*
- *A Bibliometric-Based Assessment of Research Activity in the People's Republic of China and Mini-Review of Condensed Matter Physics and Materials Science Research in the People's Republic of China*
- *New Directions for U.S.-Latin America Cooperation in Science and Technology*
- *Industrial Participation in the NSF's Engineering Research Centers*
- *Forecasting Malaysia's Science and Technology Human Resources and Research and Development Investment Needs Leading to the Year 2020*
- *The Role of NSF's Engineering Support in Enabling Technological Innovation.*

MCC GLOBAL TECHNOLOGY SERVICES

The following are recent reports from MCC:

- *Overview of Packaging Trends in Recent Japanese Portable Electronics*
- *Updates on High Density Packaging in Japan*
- *Overview of Secondary Battery Market*
- *Overview of Japanese Battery Companies*
- *Overview of Flat Panel Display Market*
- *Software Engineering and Software Process Improvement, Japan (STT Report)*
- *Mobile Telecommunications / Wireless Network Services, Europe*

- *Mobile Telecommunications Nomadic Computing, Asia*
- Strategic Technology Tour technology reports.

ONR ASIA REPORTS

The following is a small sample of ONR Asia reports with their dates. Visit **http://www.onr.navy.mil/onrasia** for a full list.

- *Some MITI (AIST) Nanoscience Plans, '92–'97,* V. Rehn, February 5, 1993
- *R&D Association for Future Electron Devices,* V. Rehn, June 1995
- *Electro-Optics Research in India,* April 15, 1996
- *Functional Polymers Research in China,* August 18–September. 8, 1996
- *Membrane Science and Technology Program at New South Wales University,* December 13, 1996
- *Ceramics Research in Asia,* December 9, 1997
- *Current Trends of Polymer Research in India,* May 1, 1998
- *Trends of Composites Research in Japan,* May 8, 1998
- *Magnetics Research in Bangladesh,* June 8, 1998
- *Electroluminescence Research in Japan,* June 25, 1998.

ONR EUROPE NEWSLETTERS

This following is a small sample of ONR Europe Newsletters, with the type and number in parentheses. Visit **http://www.ehis.navy.mil** for a full list.

- "Maritime Research Institute (MARIN), The Netherlands: An Update" (*Hydrodynamics & Acoustics Newsletter,* No. 1)
- "Acoustic Research in Russia: A Review" (*Biophysics Newsletter,* No. 41)
- "A Visit of Italy: Advances in Ship Technologies" (*Hull Machinery & Electrical Systems,* 97-002)
- "Thinking About Technology: The European Union's Institute for Prospective Technological Studies" (*Science & Technology Policy Newsletter,* 97-005)
- "Materials Research in Norway" (*Materials Science and Engineering Newsletter,* 97-004)
- "Rapid Prototyping and Composite" (PMC) Research at the University of Nottingham (*Materials Science and Engineering Newsletter,* 97-003)
- Report on Royal Institute of Technology" [Kungl Tekniska Högskolan](KTH) (*Adjunct Scientist Newsletter,* No. 3).

ATIP PUBLIC REPORTS

The following is a small sample of public ATIP reports, with either the report number or the date. Visit **http://www.atip.org** for a full list.

- *Fiber-Reinforced Elastomers—Flexible Composites in Japan*, ATIP98.001
- *Optical MEMS in Asia*, ATIP98.002
- *Taiwan's Aerospace Industries—Part I*, ATIP98.003
- *Korea's Tribology Research Center (TRC)*, ATIP98.006
- *MEMS Development in Taiwan*, ATIP98.007
- *Computer Security in Japan Status Report*, ATIP98.008
- *Venture Capital (VC) in Taiwan*, ATIP98.009
- *Technology Help to Korean Small-Medium Enterprises*, ATIP98.010
- *Asia Motor Company's Diesel-Electric Hybrid Bus (Korea)*, ATIP97.012
- *Current State of Superconductivity R&D in Japan*, ATIP97.015
- *Photonics Taiwan '96*, ATIP97.016
- *Singapore's Smart Taxis*, ATIP96.016
- *CD ROM Trends: US and Japan Compared*, ATIP95.84
- *Economic/Industrial Overview of Northeast China*, ATIP95.86
- *China's Science Parks*, ATIP95.89
- *Indian Parallel Processing Activities*, April 25, 1994
- *Nissan Motors Computer Modeling*, April 25, 1994
- *Micromachine Technology Center & National Project*, May 28,1993
- *Fifth Generation Computer System (FGCS) Summary*, June 1, 1993
- *Supercomputers in Japan*, February 7,1992
- *Parallel Processing Research in Japan*, Supplement, May 30, 1991
- *Science and Technology Research in Japan's Future*, December 7, 1990.

Ailes, Catherine P., "Assessment of Audiences for International S&T Information," Arlington, Va.: SRI International, March 1991.

Alic, John A., et al., *Beyond Spinoff: Military and Commercial Technologies in a Changing World*, Cambridge, Mass.: President and Fellows of Harvard College, 1992.

Branscomb, Lewis, ed., *Empowering Technology*, Cambridge, Mass.: MIT Press, September 1995.

Bush, Vannevar, *Science: The Endless Frontier*, reprint, Washington, D.C.: National Science Foundation, 1960.

Callan, Bénédicte, Sean S. Costigan, and Kenneth H. Keller, "Exporting U.S. High Tech: Fact and Fiction About the Globalization of Industrial R&D," New York: Council on Foreign Relations, 1999. Available at **http://www.foreignrelations.org/public/pubs/global_141.html** (last accessed July 13, 1999).

Carnegie—*see* the Carnegie Commission on Science, Technology, and Government.

The Carnegie Commission on Science, Technology, and Government, *Science and Technology in U.S. International Affairs*, New York: The Carnegie Commission on Science, Technology, and Government, January 1992, p. 40.

Council on Competitiveness, *Going Global*, Washington, D.C., 1998.

Clark, Robert M., "Scientific and Technical Intelligence Analysis," in H. Bradford Westerfield, ed., *Inside CIA's Private World*, New Haven, Conn.: Yale University Press, 1995, pp. 293–297.

Florida, Richard, "Other Countries' Money," *Technology Review*, March/April 1998, p. 31.

Gassman, Oliver, and Maximilian von Zedtwitz, "Organization of Industrial R&D on a Global Scale," *R&D Management*, Vol. 28, No. 3, July 1998, p. 131.

Hill, Christopher T., Senior Specialist, Science & Technology, Japanese *Techni-cal Information: Opportunities to Improve U.S. Access*, Washington, D.C.: Congressional Research Service, 87-818 S, October 13, 1987.

Janes, M. C., "A Review of the Development of Technology Assessment," *International Journal Technology Management, Special Issue on Technology Assessment,* Vol. 11, Nos. 5/6, 1996, pp. 507–522.

Jankowski, John E., "R&D: Foundation for Innovation," *Research Technology Management,* Vol. 41, No. 2, March-April 1998, p 14.

Jones, R.V., *The Wizard War: British Scientific Intelligence 1939-1945,* New York: Coward, McCann & Geoghegan, Inc., 1978.

Kamien, Morton I., and Nancy L. Schwartz, *Market Structure and Innovation,* Cambridge, U.K.: Cambridge University Press, 1982.

Keatley, Anne G., ed., *Technological Frontiers and Foreign Relations,* National Academy of Sciences, National Academy of Engineering, Council on Foreign Relations, Washington, D.C.: National Academy Press, 1985.

Kostoff, R. N., "Peer Review: The Appropriate GPRA Metric for Research," *Science Magazine,* Vol. 277, No. 5326, August 1, 1997, pp. 651–652.

_____, "Handbook of Research Impact Assessment," 7th ed., DTIC Report Number ADA296021, Summer 1997. Available at http://www.dtic.mil/dtic/kostoff/index.html (last accessed July 13, 1999).

Marien, Bruce A. "Protecting Scientific Excellence" *Research Technology Management,* Vol. 41, No. 2, March–April 1998, p. 39.

McLoughlin, Glen, *International Science and Technology Issues for U.S. Policy-makers,* Washington, D.C.: Congressional Research Service, 94-733 SPR, September 16, 1994.

National Economic Council, National Security Council, and Office of Science and Technology Policy, *Second to None: Preserving America's Military Advantage Through Dual-Use Technology,* Ft. Belvoir, Va.: Defense Technical Information Center, ADA 286779, February 1995.

National Science Board, *Science and Engineering Indicators-1998,* Washington, D.C.: National Science Foundation, NSB-98-1, 1998.

National Science Board, Committee on Industrial Support for R&D, National Science Foundation, *The Competitive Strength of U.S. Industrial Science and Technology: Strategic Issues,* Washington, D.C.: U.S. Government Printing Office, NSB-92-138, August 1992.

NSB—*see* National Science Board.

Ohmae, Kenichi, *The Borderless World: Power and Strategy in the Interlinked Economy*, New York: Harper Business, 1990.

Popper, Steven, Caroline Wagner, and Eric Larson, *New Forces at Work: Industry Views Critical Technologies*, Santa Monica, Calif.: RAND, MR-1008-OSTP, 1998.

Prestowitz, Clyde V., Jr., *Trading Places: How We Allowed Japan to Take the Lead*, New York: Basic Books, 1988.

Ramo, Simon, "The Foreign Dimension of National Technological Policy," Simon Ramo, in *Technological Frontiers and Foreign Relations*, Washington, D.C.: National Academy Press, 1985, p. 14.

Reich, Robert, and Ira Magaziner, *Minding America's Business*, New York: Vintage Books, 1982.

Roosevelt, President Franklin, letter to Vannevar Bush, November 17, 1944, reprinted in Bush (1960).

Rosenberg, Nathan, *Inside the Black Box*, Cambridge, U.K.: Cambridge University Press, 1982.

Smith, Bruce L. R., and Claude E. Barfield, eds., *Technology, R&D, and the Economy*, Washington, D.C.: The Brookings Institution and the American Enterprise Institute for Public Policy Research, 1996.

U.S. Department of Commerce, Office of Technology Policy, *Foreign Science & Technology Information Sources: In the Federal Government and Select Private Sector Organizations*, Washington, D.C., 1996

U.S. General Accounting Office, *Foreign Technology: U.S. Monitoring and Dissemination of the Results of Foreign Research*, GAO Report to the Chairman, Subcommittee on Technology and National Security, Joint Economic Committee, U.S. Congress, Washington, D.C., GAO/NSAID-90-117, March 1990.

U.S. General Accounting Office, *Foreign Technology: Federal Processes for Collection and Dissemination*, GAO Report to the Chairman, Subcommittee on Defense Industry and Technology Committee on Armed Services, U.S. Senate, Washington, D.C., GAO-NSAID-92-101, March 1992.

U.S. General Accounting Office, *Foreign Technology: Collection and Dissemination of Japanese Information Can Be Improved*, GAO Report to the Chairman, Subcommittee on Defense Technology, Acquisition, and Industrial Base, Committee on Armed Services, U.S. Senate, Washington, D.C., GAO/NSAID-93-251, September 1993.

U.S. Congress, Office of Technology Assessment, *Helping America Compete: The Role of Federal Scientific & Technical Information*, Washington, D.C.:

U.S. Government Printing Office, OTA-CIT-454, June 1990. Also available via **http://www.wws.princeton.edu/~ota** (last accessed July 2, 1999).

Wagner, Caroline S., *Techniques and Methods for Assessing the International Standing of U.S. Science*, Santa Monica, Calif.: RAND, MR-706.0-OSTP, October, 1995.

The White House, *Technology for America's Economic Growth*, Washington, D.C., February 22, 1993.

The White House, *Science in the National Interest*, Washington, D.C., August 3, 1994.

World Technology Evaluation Center, *Electronics in the Pacific Rim*, Baltimore, Md.: International Technology Research Institute, Loyola College, 1997.

Woolsey, R. James, Director of Central Intelligence, quoted in the *Journal of Electronic Defense*, January 1998, p. 30.